CHEMICAL ANALYSIS

Other volumes in preparation

CHEMICAL ANALYSIS

A SERIES OF MONOGRAPHS ON

ANALYTICAL CHEMISTRY AND ITS APPLICATIONS

Editors

P. J. ELVING • I. M. KOLTHOFF

VOLUME 30

WILEY-INTERSCIENCE

A Division of John Wiley & Sons,
New York/London/Sydney/Toronto

ULTRAMICRO
ELEMENTAL ANALYSIS

by

GÜNTHER ⌐TÖLG

translated by

CONRAD E. THALMAYER

Library of Congress Catalogue Card Number: 71-105391

SBN 471 87675 5

Printed in the United States of America

10 9 8 7 6 5 4 3 2 1

To the memory of

Professor Wilhelm Geilmann

PREFACE

Procedures for organic elemental analysis in the range of 30–100 μg were first developed fifteen years ago by R. Belcher and his school in Birmingham, England. This pioneering work has stimulated the development in many places of new procedures with the goal of simplifying the methods or using even smaller amounts of sample. At present there is only one monograph, in which Professor Belcher has reported on his own procedures. It seemed fitting, therefore, to compile all the available work in this area. In so doing, it was possible to review my own work with very small samples over the previous ten years.

For his helpful interest in our work, I thank Professor Belcher sincerely.

I am especially grateful to Professor R. Bock, Mainz, who not only was actively concerned with the writing of this book, but also was always helpful with good suggestions and valuable criticism during the development of our methods.

Furthermore, I cordially thank Professor R. Neeb, Dr. K. Ballschmiter, Dr. K. Behrends, Mr. W. H. List, Mr. A. Grünert, and Mr. I. Schuphan for their critical inspection of the manuscript and their help in proofreading.

GÜNTHER TÖLG

Mainz, Germany, autumn 1967

vii

CONTENTS

1. GENERAL

1.1. INTRODUCTION

In the first third of this century, F. Pregl and his school showed that the concentration of an element in an organic compound can be determined with a few milligrams of sample if suitably refined apparatus and, above all, capable analytical balances are used. This "organic elemental microanalysis" still satisfies most requirements with respect to accuracy and sample size; nearly all improvements in this field affect only efficiency. A decrease in sample size yields no advantages, the difficulty increases, and sufficient sample homogeneity is often no longer assured. The necessity of developing microgram, or even nanogram, procedures appears more and more in connection with biological, biochemical, and medical problems and in the elucidation of natural substances. Often only a very small amount of sample is available for analysis, so that the greater difficulty of ultramicroprocedures must then be accepted.

While numerous methods have been developed (1–31) for the determination of the individual elements in the region of 100–1000 μg sample size (decimilligram region), there has been relatively little experience with procedures which require smaller samples (centimilligram region and below).

On the basis of statistical considerations, the minimum required sample size is about 10^{-16} g if the accuracies of microprocedures are to be attained (32,33). The amount actually needed depends primarily on the properties of the substance to be analyzed (homogeneity, stability, volatility, solubility), on the accuracy of the sample weight, and on whether the small sample can still be manipulated reliably.

There are two possible approaches to the determination:

1. The decomposition methods developed for the milligram or decimilligram region are adapted to the smaller samples. Unless the reaction products can be determined by gas chromatography, gas

1

volumetry or manometrically, as in the case of carbon and hydrogen determination, it is preferable to determine them titrimetrically or spectrophotometrically.

The concentration relationships found convenient in micro-procedures are maintained by scaling down the apparatus and the quantity of reagents. In the extreme case this results in working in fine capillaries with only a few microliters of solution (33–37); in this case only the relative increase in surface area of the container and the associated effects (increase of adsorption and desorption) and the increase in the rate of evaporation are pertinent.

Assuming solution volumes of 0.5–2 ml, which can still be conveniently manipulated, and titrant solutions of normality not below 0.01, as required in customary microanalysis (38), then about 30–100 μg sample is required for the determination of an element. This region, in which it is still possible to weigh the sample directly with sufficient accuracy, was made accessible to organic elemental analysis in recent years primarily by Belcher and his co-workers (39–42). The procedures which have so far been worked out for samples of this size are summarized in Table 1.

Table 1. Procedures for Samples of 20–100 μg

Element	Sample weight, μg	Method of decomposition	Method of determination	References
C	>30	O_2/Pt/MnO_2	CO_2, manometric	44
		O_2	CO_2, Hg displacement, gravimetric	45
C, H	>30	O_2/Pt/silver manganate	CO_2 and H_2O, manometric	46
C, H, N	>40	CuO/He	Thermal conductivity measurement	74
N	50–100	H_2SO_4/Hg (II) salt	NH_3, titrimetric with NaOBr	47, 48
	>50	H_2SO_4	NH_3, turbidimetric as NH_4–tetraphenyl-borate	49
	30–100	O_2/Cu	N_2, Hg displacement, gravimetric	50
	>30	Sealed tube combustion	Manometric	75
O	20–60	After Unterzaucher	Iodometric	51

Table 1   (continued)

Element	Sample weight, μg	Method of decomposition	Method of determination	References
S	30–100	After Carius or in the oxygen flask	Sulfate, titrimetric	52, 53
	50–100	After Carius	Sulfate, titrimetric	54
	>60	Reduction with Sn(II) phosphate in H_3PO_4	Sulfide, titrimetric	76
F	30–80	In the oxygen flask	Spectrophotometric with Ce (III) or La–alizarin fluorine blue	55–57
Cl	30–80	In the oxygen flask	Mercurimetric titration	58
Br	30–90	With Na or in the oxygen flask	BrO_3^-, iodometric	59, 60
	30–100	In the oxygen flask	Spectrophotometric	61
I	30–70	In the oxygen flask	IO_3^-, iodometric	59, 60
	>12	In the oxygen flask	Argentometric titration	77
P	30–100	$HClO_4/HNO_3$ or in the oxygen flask	Spectrophotometric or titrimetric	62-64
	>40	Open tube digestion with $HClO_4/H_2SO_4$	Spectrophotometric	78
As	30–100	In the oxygen flask	Spectrophotometric	64
	>15	In the oxygen flask	Titrimetric	79
Fe	30–100	After Carius	Spectrophotometric with 1,10-phenanthroline	65

Table 2. Results of Procedures with Samples Under 30 μg

Element	Sample weight, μg	Method of decomposition	Method of determination	Standard deviation, μg	References
C	10–30	O_2	CO_2, complexometric–spectrophotometric titration	±0.06	66
H	10–30	O_2	H_2O, after conversion to H_2S by argentometric–bipotentiometric titration	±0.01	67
N	5–20	Hydrogenating digestion	NH_3, iodometric–biamperometric titration	±0.01$_6$	68
S	2–10	O_2, H_2O_2	SO_4^{2-}, by preciptation–titration, visual	±0.01	69
		Hydrogenating digestion	H_2S, titration with HBrO, biamperometric		43
F	2–10	In the oxygen flask	Spectrophotometric	±0.005	69
Cl	2–10	O_2	Cl^-, argentometric–bipotentiometric	±0.01$_5$	70
Br	2–10	Hydrogenating digestion	Br^-, argentometric–bipotentiometric	0.01	71
I	1–5	In the oxygen flask	I^-, argentometric, iodometric, or spectrophotometric	0.005	72
P	1–5	$HClO_4/HNO_3$ or in the oxygen flask	PO_4^{3-}, spectrophotometric	0.003	69

2. A further reduction of sample size leads, if volumes of 0.5 to about 2 ml are to be retained in the determination, to the concentrations usual in trace analysis. Additional sources of error now appear, primarily by greater exchange of reaction products with the walls of the apparatus and by irreproducible blanks. Nevertheless, it has been shown (43, 80) that sufficient accuracy can be realized (cf. Table 2).

Since samples under about 20 μg with tare in the milligram region can at present be accurately measured out only by solution partition, these procedures are limited to soluble substances of low volatility. Statistical and systematic errors remain small if the utensils have as small a surface area as possible and are of suitable materials (e.g., quartz, noble metals, Teflon, polyethylene, polypropylene), only small amounts of easily purified reagents are used, and the introduction of impurities, especially from the laboratory air, is minimized. Since "personal error" can make up a large part of the total error in ultramicroprocedures, good reproducibility is achieved only if a procedure is composed of as few constituent operations as possible which take place under experimental conditions that are maintained constant.

Of all the analytical procedures examined, titrimetric procedures have shown themselves to be the most accurate. Titration procedures with visual endpoint indication are useful only under certain conditions (69,72) for the determination of sample concentrations under 1 μg per milliliter solution volume, but good accuracy with even 10^{-3} to 10^{-4} N solutions is obtained if the titrations are carried out in organic solvents and the endpoints are indicated electrochemically or photometrically. These procedures can be automated by the application of motor-driven micro piston burets or coulometric methods (73).

Spectrophotometric procedures, too, can be sufficiently accurate for the determination of very small amounts of elements, such as phosphorus, fluorine, and iodine, which are less susceptible to impurities that may be introduced. There are still, however, only few reactions suitable that lead to stable, deeply colored compounds that are extractible with organic solvents. Absorbance measurements in the ultraviolet region are generally more subject to interference than measurements in the visible region. Analysis volumes have been chosen no smaller than 1 ml, so that spectrophotometers without microcuvets can be used; further increases in sensitivity are possible by the use of capillary cuvets (80).

1.1.1. REFERENCES

1. M. Williams, in I. M. Kolthoff and P. J. Elving, *Treatise on Analytical Chemistry*, Part II, Vol. 11, Interscience, New York, 1965.
2. Al Steyermark, *Quantitative Organic Microanalysis*, Academic Press, New York, 1965.
3. R. Levy, *"Microanalyse Organique Elementaire,"* in *Monographies de Chimie Organique*, Masson, Paris, 1961; *Chim. Anal.*, **46**, 113 (1964).
4. T. S. Ma and M. Gutterson, *Anal. Chem.* **38**, 186 R (1966).
5. W. J. Kirsten, *Z. anal. Chem.*, **181**, 1 (1961).
6. W. Merz and W. Pfab, *Microchem. J.*, **10**, 346 (1966).
7. W. Walisch, *Ber. Deut. Chem. Ges.*, **94**, 2314 (1961).
8. H. Weitkamp and F. Korte, *Chem.-Ing.-Tech.*, **6**, 429 (1963).
9. F. Ehrenberger, H. Kelker, and O. Weber, *Z. Anal. Chem.*, **222**, 260 (1966).
10. G. Kainz and E. Wachberger, *Z. Anal. Chem.*, **222**, 271 (1966).
11. C. D. Miller and J. D. Winefordner, *Microchem. J.*, **8**, 334 (1964).
12. C. D. Miller, *Microchem. J.*, **11**, 366 (1966).
13. W. Simon, P. F. Sommer, and G. H. Lyssy, *Microchem. J.*, **6**, 239 (1962).
14. H. Simon and G. Müllhofer, *Z. Anal. Chem.*, **181**, 85 (1961).
15. C. W. Koch and E. E. Jones, *Mikrochim. Acta*, **1963**, 734.
16. F. Martin, A. Floret, and J. Lemaitre, *Bull. Soc. Chim. France*, **1964**, 1836
17. W. J. Kirsten, K. Hozumi, and L. Nirk, *Z. Anal. Chem.*, **191**, 161 (1962).
18. E. Maly and J. Krsek, *Mikrochim. Acta*, **1964**, 778.
19. K. Yoshikawa and T. Mitsui, *Microchem. J.*, **9**, 52 (1965).
20. R. N. Boos, *Microchem. J.*, **8**, 389 (1964).
21. G. Kainz and E. Wachberger, *Z. Anal. Chem.*, **222**, 278 (1966).
22. K. Hozumi, *Anal. Chem.*, **38**, 641 (1966).
23. A. Pietrogrande, *Mikrochim. Acta*, **1964**, 1106.
24. I. E. Pakhomova and M. N. Chumachenko, *Izv. Akad. Nauk SSSR, Ser. Khim.*, **1965**, 1138.
25. H. Dirscherl and F. Erne, *Mikrochim. Acta*, **1963**, 242.
26. J. E. Fildes and W. J. Kirsten, *Microchem. J.*, **9**, 411 (1965).
27. E. Pell, L. Machherndl, and H. Malissa; *Microchem. J.*, **10**, 286 (1966).

28. W. J. Kirsten, *Microchem. J.*, **7**, 34 (1963).
29. W. J. Kirsten, B. Danielson, and E. Öhren, *Microchem. J.*, **12**, 307 (1967).
30. A. Campiglio, *Farmaco, Ed. Sci.*, **19**, 1033 (1964).
31. T. Onoe and H. Shimada, *Rep. Sankyo Res. Lab.*, **17**, 82 (1965).
32. A. A. Benedetti-Pichler, *Mikrochim. Acta*, **36/37**, 38 (1951).
33. I. P. Alimarin and M. N. Petrikova, *Anorganische Ultramikroanalyse*, Deutscher Verlag der Wissenschaften, Berlin, 1962.
34. I. M. Korenman, *Introduction to Quantitative Ultramicroanalysis*, Academic Press, New York, 1965.
35. H. Mattenheimer, *Mikromethoden für das klinisch-chemische und biochemische Laboratorium*, 2nd ed., de Gruyter, Berlin, 1966.
36. P. L. Kirk, *Quantitative Ultramicroanalysis*, 2nd ed., Wiley, New York, 1951.
37. J. S. Wiberley, *Microchem. J.*, **11**, 343 (1966).
38. J. Mika, *Die Methoden der Mikromassanalyse, Die chem. Analyse*, Vol. 42, Ferd. Enke-Verlag, Stuttgart, 1958.
39. R. Belcher, *Submicro Methods of Organic Analysis*, Elsevier, Amsterdam, 1966.
40. R. Belcher, *Chem.-Ing.-Tech.*, **36**, 993 (1964).
41. R. Belcher, in *Microchemical Techniques*, Vol. 2, N. Cheronis, Ed., Wiley, New York, 1962, pp. 107–115.
42. R. Belcher, *Z. Anal. Chem.*, **181**, 22 (1961).
43. G. Tölg, Habilitationsschrift, Mainz, 1965.
44. C. W. Ayers, R. Belcher, and T. S. West, *J. Chem. Soc.*, **1959**, 2582.
45. W. J. Kirsten and K. Hozumi, *Mikrochism. Acta*, **1962**, 777.
46. P. Gouverneur, H. C. E. van Leuven, R. Belcher, and A. M. G. Macdonald, *Anal. Chim. Acta*, **33**, 360 (1965).
47. R. Belcher, T. S. West, and M. Williams, *J. Chem. Soc.*, **1957**, 4323.
48. R. Belcher, A. D. Campbell, and P. Gouverneur, *J. Chem. Soc.*, **1963**, 531.
49. R. A. Shah and N. Bhatty, *Mikrochim. Acta*, **1967**, 81.
50. K. Hozumi and W. J. Kirsten, *Anal. Chem.*, **34**, 434 (1962).
51. A. Campiglio, A. *Mikrochim. Acta*, **1964**, 114.
52. R. Belcher, R. L. Bhasin, R. A. Shah, and T. S. West, *J. Chem. Soc.*, **1958**, 4054.
53. R. Belcher, P. Gouverneur, A. D. Campbell, and A. M. G. Macdonald, *J. Chem. Soc.* **1962**, 3033.
54. D. C. White, *Mikrochim. Acta*, **1959**, 254; **1962**, 807.

55. R. Belcher, M. A. Leonard, and T. S. West, *J. Chem. Soc.,* **1959**, 3577.
56. M. E. Fernandopulle and A. M. G. Macdonald, *Microchem. J.,* **11**, 41 (1966).
57. T. R. F. W. Fennell and J. R. Webb, *Microchem. J.,* **10**, 456 (1966).
58. R. Belcher, P. Gouverneur, and A. M. G. Macdonald, *J. Chem. Soc.,* **1962**, 1938.
59. R. Belcher, R. A. Shah, and T. S. West, *J. Chem. Soc.,* **1958**, 2998.
60. R. Belcher, Y. A. Gawargious, P. Gouverneur, and A. M. G. Macdonald, *J. Chem. Soc.,* **1964**, 3560.
61. T. R. F. W. Fennell and J. R. Webb, *Z. Anal. Chem.,* **205**, 90 (1964).
62. W. J. Kirsten and M. E. Carlsson, *Microchem. J.,* **4**, 3 (1960).
63. T. Salvage and J. P. Dixon, *Analyst,* **90**, 24 (1965).
64. R. Belcher, A. M. G. Macdonald, S. E. Phang, and T. S. West, *J. Chem. Soc.* **1965**, 2044.
65. T. R. F. W. Fennell and J. R. Webb, *Talanta,* **9**, 795 (1962).
66. K. H. Ballschmiter and G. Tölg, *Z. Anal. Chem.,* **203**, 20 (1964).
67. G. Tölg and K. H. Ballschmiter, *Microchem. J.,* **9**, 257 (1965).
68. G. Tölg, *Z. Anal. Chem.,* **205**, 40 (1964).
69. G. Tölg, *Z. Anal. Chem.,* **194**, 20 (1963).
70. G. Schwab and G. Tölg, *Z. Analyt. Chem.,* **205**, 29 (1964).
71. W. H. List and G. Tölg, *Z. Analyt. Chem.* **226**, 127 (1967).
72. B. Morsches and G. Tölg, *Z. Analyt. Chem.,* **200**, 20 (1964).
73. K. Abresch and J. Claassen, *Die coulometrische Analyse,* Verlag Chemie, Weinheim/Bergstr. 1961, p. 131.
74. R. Belcher, G. Dryhurst, A. M. G. Macdonald, J. R. Majer and G. J. Roberts, *Anal. Chim. Acta,* **43**, 441 (1968).
75. K. Hozumi and K. Umemoto, *Microchem. J.,* **12**, 512 (1967).
76. B. Griepink, J. Slanina and J. Schoonman, *Mikrochim. Acta,* **1967**, 984.
77. B. Griepink and A. van Sandwijk, *Mikrochim. Acta,* **1969**, 1014.
78. A. J. Christopher and T. R. F. W. Fennel, *Microchem. J.,* **12**, 593, (1967).
79. B. Griepink, W. Krijgsman, A. J. M. E. Leenaers-Smeets, J. Slanina and H. Cuijpers, *Mikrochim. Acta,* **1969**, 1018.
80. G. Tölg and I. Lorenz, "Methoden der mikrochemischen Elementbestimmung und ihre Grenzen," in *Fortschritte der chemischen Forschung,* Vol. 11/4, Springer-Verlag, Berlin-Heidelberg-New York, 1969.

1.2. MEASUREMENT AND MANIPULATION OF ULTRAMICRO SAMPLES

1.2.1. WEIGHING

1.2.1.1. ULTRAMICROBALANCES

Gravimetric determination methods are of no significance in organic microgram analysis. The balance now serves only for weighing out the materials to be examined. Extremely sensitive ultramicrobalances, which have been constructed according to various principles with accuracies of less than 10 ng (1–4), possess only very small capacities and are very susceptible to vibration, temperature fluctuations, dust, and air-borne impurities. Moreover, the manipulation of samples under about 20 μg requires optical aids and mechanical manipulators (5). Somewhat larger samples, on the other hand, can be weighed sufficiently accurately and quickly with quartz fiber torsion balances. With capacities of 200–400 mg, these possess absolute accuracies between 10 and 50 ng and are relatively insensitive to vibration and temperature fluctuations (cf. Sect. 1.2.2.1).

The electric ultramicrobalances currently commercially available, in which the change of equilibrium position of the beam produced by a load is electromagnetically compensated, are too inaccurate[1] for samples of less than 100 μg.

For ultramicro beam balances[2] too, reproducibilities of only ±0.1 to ±0.2 μg (difference weighings) are specified for the weighing region of 0–2 mg.

1.2.1.2. INDIRECT METHODS

Samples of more than 1 μg can be measured out with a relative standard deviation of about ±0.2% by solution partition (6). In

---

[1] Thus, for example, the Cahn Gram Electrobalance (Cahn Instrument Co., Paramount, Calif.) has an accuracy of about ±0.1 μg in the 0–1000 μg weighing region (prospectus specification) and the Mark II Balance Model A (C. I. Electronics Ltd., Staines, Middlesex, England) has an accuracy of only ±0.5 μg in the most sensitive weighing region of 0–25 μg (prospectus specification).

[2] For example, Type UM7 of Mettler Co., Zurich, or Type 25 UM of Bunge Co., Hamburg.

this technique, a sample which can still be weighed out accurately to ±0.1% with an ultramicrobalance is dissolved in a known volume of suitable solvent. For the determination of the several elements of a compound and for parallel determinations, aliquot portions are employed in which about 20 μl can be measured out with a relative standard deviation of ±0.1%.

Thus, for example, a sample of 50 ±0.05 μg yields, after solution in 200 μl solvent, 8–9 aliquot portions of 5 μg each. The procedure is described exhaustively in Sect. 1.2.4.4.

Prerequisites for a solution partition are that the sample be nonvolatile and soluble in a relatively volatile solvent, so that no losses can occur upon removal of the solvent.

These restrictions clearly show the expedient on which we are dependent as long as direct, accurate weighing of samples of less than 20 μg with tare in the milligram region engenders difficulties. Yet in many cases this technique does make it possible to carry out a complete elemental analysis with samples of between 50 and 100 μg.

Other indirect, sensitive methods of weight determination by way of measurement of volume and density of the substance to be examined (7,8) are meaningless for organic elemental analysis.

1.2.2. WEIGHING WITH QUARTZ FIBER TORSION BALANCES

1.2.2.1. PRINCIPLE

Quartz fiber torsion balances (cf. Figs. 1,2) are built with two arms. A quartz beam of about 200–300 μm diameter with suspensions of fine quartz fibers is fused or cemented onto a very thin quartz fiber (10–50 μm). This quartz fiber is fastened rigidly at one end to the balance case; at its other end it is fastened to the shaft of a vertical torsion wheel with a graduated circumference. When the beam has been brought out of its equilibrium position by being overloaded on one side by the mass m, it can be restored, by rotation of the torsion wheel in the opposite direction, to its original

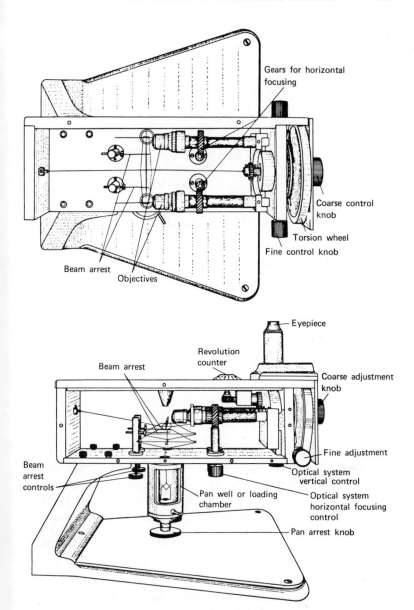

Figure 1. Rodder Ultramicrobalance, Model E.

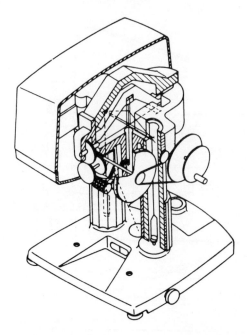

Figure 2. Quartz fiber torsion balance Q01
of Oertling Ltd., England.

position, which is observed with graduated telescopes or light spots.

The torque necessary to compensate for the excess weight is

$$T = m \cdot g \cdot l_B = \frac{b \cdot r^4 \cdot \tau}{l_F}$$

where g represents the acceleration due to gravity, l_B the length
of the beam from the quartz fiber to the suspension support, b the
arc read off the graduated circumference of the torsion wheel, r the
radius, l_F the length of the quartz fiber, and τ the coefficient of
torsion of quartz. Here r and l_F refer only to the portion of the
torsion fiber toward the graduated wheel.

The balance is more sensitive the thinner and longer the quartz
fiber. Since the torque depends on the fourth power of the fiber
radius, the sensitivity grows far more rapidly with decreasing r
than the capacity drops. The latter is proportional to the square
of r. In highly sensitive balances the center of gravity and the ful-

crum are as close together as possible; furthermore, the fulcrum should be in the plane of the suspension supports.

The length of the quartz fiber between torsion wheel and beam determines the weighing range, for a longer fiber can be twisted more.

In Table 3 the most important types of quartz fiber torsion balances and their properties are summarized.

Exhaustive descriptions are given by Benedetti-Pichler (1). Gorbach (9,10), Behrndt (11), and Ingram (12) give short surveys.

1.2.2.2. SETTING UP

Because of their strong damping, torsion balances are relatively insensitive to vibration; thus, unless extreme sensitivity is to be attained, they can be placed on ordinary, firm tables. At times, it is true, periodic vibrations (e.g., of motors) can come into resonance with the oscillating system of the balance, thus strongly impairing the accuracy of the weighing.

Because of the very small coefficient of thermal expansion of quartz, slow variations in room temperature of a few degrees have only an insignificant effect, but rapid temperature changes (e.g., caused by exposure to variable sunshine) and uneven heating, in spite of the good insulation of the commercial balances, must absolutely be avoided. The normal variations in atmospheric humidity are unimportant.

To eliminate variable electrostatic charges from the beam and the case, which prevent the beam from swinging freely upon release, the case should be grounded and a source of weak β radiation (e.g., ^{90}Sr) placed in the balance to ionize the air.

Since dust particles can weigh up to about 1 μg, it is necessary to work absolutely dustfree, either by using dustfree weighing rooms or by setting up the balance in a glove box. The second method is simpler and more efficient.

Figure 3 shows a Plexiglas design which makes it possible to conduct preparations and weighings with dust excluded. The ultramicrobalance is in the left chamber (width: 500 mm; depth: 670 mm; height: 700 mm). This weighing chamber is connected to the center preparation chamber (width: 1000 mm; depth: 670 mm;

Table 3. Summary of Quartz Fiber Torsion Balances

Author	Capacity of each pan, mg	Absolute accuracy, ng	Remarks
Neher (13)	1	1	First quartz fiber torsion balance.
Kirk, Craig, Gullberg, and Boyer (14)[a]		0.5	Stretched quartz fiber, torsion wheel with vernier scale, bending support.
El-Badry and Wilson (15)	200	40	As above, except the quartz parts are not fused but cemented.
Garner (16)[b]	5000	100	Relief of the torsion fiber by additional suspension of the beam.
Korenman, Fertel'meyster, and Rostokin (17)		50	Rotation of the torsion wheel is geared down.
Carmichael (18)	300	10	
Rodder (19)[e]	4	1	Fiber is not
(Model E)	200	$\pm 30^e$	stretched; special end supports with double suspension; torsion wheel with vernier scale.
Asbury, Belcher, and West (20)[d]	250	$\pm 40^e$	Rotation of the torsion wheel is geared down; projected scale.

[a] Microchemical Specialties Co., Berkeley 3, Calif.

[b] Vortox Co., Claremont, Calif.

[e] Microtech. Services Co., Los Altos, Calif.

[d] L. Oertling Ltd., Cray Vakey Works, St. Mary Cray, Orpington, Kent, England.

[e] Absolute standard deviation (50 weighings of a platinum wire of ca. 50 μg) according to measurements of the author.

height: 700 mm), which is provided with rubber gloves and an air lock. The rear wall is detachable so that large devices can also be set up inside the box. Electricity, gases, and water are admitted

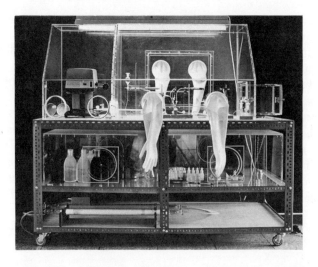

Figure 3. Movable glove box for weighing and handling small samples. Upper left: balance chamber with quartz fiber torsion balance; center: preparation chamber with gloves; right: air lock. Center compartment: four separate Plexiglas chambers for dust-proof storage of reagents and apparatus. Lower compartment: air purification train, consisting of two glass tubes (1 m x 4 cm) in series, with $\overline{\mathrm{S}}$ 29 joints and stopcocks. One tube is filled with NaOH pellets, the other with ignited quartz wool.

through the floor of the box. Weighing vessels containing samples can be securely moved from the preparation chamber to the weighing chamber and back in little quartz containers lying on a Plexiglas sled which runs in a channel. Temperature and humidity in such boxes can be kept substantially constant.

1.2.2.3. USE

Weighing with quartz fiber torsion balances is simpler than with the usual beam microbalances. Because of the fragility of the quartz fibers and the other quartz parts, however, special care is required in setting down the weighing dishes. Mechanical manipulators (15) can be useful for this. In the analysis of very small samples directly weighed, it is advisable to use the weighing boat also as a container for the sample during decomposition (cf. Sect.

1.3.2); therefore the material and form of the weighing boat are generally determined by the method of decomposition.

Small platinum boats are frequently used; they can easily be made of foil in any size with suitable dies (cf. Fig. 4a). $10 \times 2 \times 2$ mm boats of 0.02-mm platinum foil weigh about 25 mg. The foil must be annealed before forming.

Quartz combustion vessels (8×2 mm) with stems (7×0.7 mm) have proved especially good for ultramicroprocedures (cf. Fig. 4c). They weigh about 40 mg. The vessels can be kept vertically or horizontally in suitably drilled aluminum receptacles. For introduction into the decomposition apparatus they are inserted into the spiral of a holder made of 0.4 mm Pt–Rh wire (cf. Fig. 4c).

To weigh solids that are to be transferred to narrow tubes or

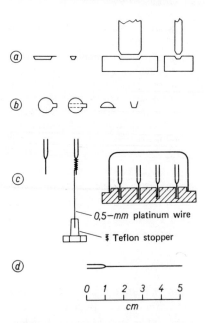

Figure 4. Various weighing vessels. (a) Platinum boat with dies. (b) Aluminum boat formed of foil. (c) Quartz vessels with holder and storage receptacle. (d) Glass weighing tube.

flasks with narrow openings, it is expedient to use weighing tubes as shown in Fig. 4d. These tubes with thin stems are prepared by drawing out melting-point tubes (6 $\times$ 1.8 mm tubes with 30 $\times$ 0.5 mm stems weigh about 25 mg).

Platinum and quartz vessels are boiled before use in semiconcentrated nitric acid, ignited, and stored protected from dust.

Aluminum boats, as used in weighing samples that are to be further divided by solution partition, are prepared from aluminum disks of equal weight (7 $\times$ 0.03 mm) with a little "nose" (cf. Fig. 4b) by forming the disks over a square peg of 2 mm width. Both ends remain open. The boats are rinsed several times with double-distilled water and pure acetone, dried at 150°, and stored protected from dust.

In accordance with principle, the counterpoise in weighing is an equally heavy vessel of the same material. For a platinum boat the weight of the counterpoise can be adjusted by careful treatment with aqua regia, for a quartz boat by treatment with hydrofluoric acid.

The boats are placed on the balance suspension with platinum-tipped forceps; during this operation the two weighing chambers should be opened for only brief periods of about equal duration. Even when the counterpoise is not being changed, its chamber should be opened for temperature and moisture equalization when the other boat is inserted. In opening and closing the chambers, their handles should be touched for only brief, equal periods. Weight differences between the boats, both in taring and in weighing, are compensated for by turning the torsion wheel until the null marks agree. The calibration is linear over the entire weighing region, i.e., knowledge of one calibration factor is sufficient, though it is not given by all manufacturers. In general the calibration factor remains constant for several months. It is essential, however, to check it from time to time and to correct it if necessary.

1.2.2.4. CALIBRATION

Several calibration methods of about equal accuracy have been described (1,2,7,21,22), of which particularly the following are recommended:

1. Analytical grade potassium chloride, dried at 200° and

weighed on a microbalance, is dissolved in a known quantity of double-distilled water. A drop of this solution is transferred from a micro weighing buret onto a piece of platinum foil and its weight determined by reweighing the buret. The drop is evaporated and the residue dried at 200°; the calculated mass of the residue and the weight of the foil with and without residue yield the calibration factor, which is reproducible to about ±0.1% (25 weighings).

2. The weight of a long piece of platinum wire of uniform diameter, thoroughly cleaned and ignited, is determined (e.g., 20.0 cm of 0.025 mm wire weigh 2.109 mg) and several pieces (e.g., 5 and 10 mm) are cut off, the lengths of which are accurately determined with the aid of a microscope with a micrometric eyepiece. Assuming that the wire is equally thick over its entire length, the weights of the pieces are calculated (e.g., 52.72 and 105.45 μg) and the calibration factor thus determined. The relative standard deviation of this method is about 0.2%.

3. A cleaned piece of platinum wire (ca. 74 × 0.04 mm, ca. 2 mg) is cut into ten pieces of different lengths and the total weight m_o of all pieces m_i (mean of at least five weighings) is determined on a suitable microbalance. Each piece is then placed on the quartz fiber torsion balance and its scale division difference on the torsion wheel d_i determined (mean of at least three weighings). From this the weight equivalent k of each scale division of the torsion wheel can be calculated according to

$$k = \frac{\Sigma m_i}{\Sigma d_i} = \frac{m_o}{\Sigma d_i} \quad (\mu g/\text{div})$$

and the weight of each platinum piece is

$$m_i = k \cdot d_i$$

The calibrating weights of different sizes thus obtained are stored in a small desiccator and are available for recalibration.

It is advisable to check the correctness of the calibration factor obtained with one procedure against that obtained by another.

1.2.3. SAMPLE PREPARATION AND WEIGHING

All determination procedures hitherto devised for samples under 100 μg assume solid samples which are nonvolatile at room temperature, nonhygroscopic, and stable, and which must be dried to constant weight before weighing. Here it should be noted that substances considered stable in weight in the milligram region may exhibit appreciable volatility in the microgram region even at room temperature, especially if the material is very finely divided and the weight loss is followed over a long time. Fennell and Webb (23) point out that to avoid sample losses one should not wait unnecessarily long between weighing and decomposition, even if no weight loss is observed in weighing. For drying the sample, vacuum drying pistols (24) suited to such small samples are used and only well filtered, dustfree, dried air should then be admitted into them. To avoid air turbulence the air must be admitted through a fine capillary, ahead of the stopcock, so that the air enters over 15–20 min. The sample should be removed only in a drybox. To remove the sample a small spatula is used (with an area of about 3 mm²), made by pounding flat and grinding the tip of a refined steel dissecting needle. The tiny sample crystals are transferred from the spatula to the weighing boat with the aid of another dissecting needle. The instruments are manipulated with hands resting on a support and observed through a lens under good illumination.

1.2.4. MEASUREMENT OF VERY SMALL AMOUNTS OF LIQUID

Of the numerous ultramicropipets and -burets described in the literature (2,3,7,25–28), the following types have especially proved themselves for the purposes of organic elemental analysis with extremely small samples.

1.2.4.1. PIPETS

Commercial glass pipets (cf. Fig. 5) are calibrated to-contain and must be rinsed out to deliver their stated volume. They have the following accuracies: for the region 1–20 μl about ±0.5%, 25–40 μl about ±0.3%, 45–85 μl about ±0.2%, and for larger volumes about ±0.1%.

To-deliver microliter glass pipets are less accurate; they have errors of about 1% and higher.

Polyethylene and polypropylene pipets (29,30) are wet far less than glass pipets by aqueous solutions and solutions in strongly polar organic solvents and can therefore also be calibrated to-deliver very

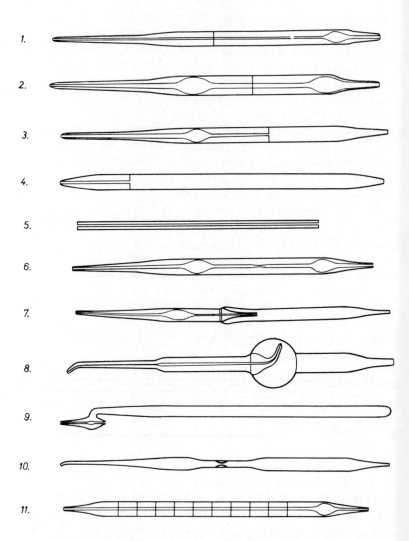

Figure 5. Conventional glass pipets for the microliter region (manufacturer: Microchemical Specialties Co., Berkeley 10, Calif.). Types calibrated to-contain: *1* and *2*, pipets after Kirk (31); *3–7*, self-filling pipets; *8*, pipet after Grunbaum-Kirk (32) with reagent reservoir; *9*, self-filling pipet. Types calibrated to-deliver: *10*, transfer pipet; *11*, measuring pipet.

accurately. If the pipets are not exposed to direct sunlight or heated over 40°, their volume remains constant for several months.

For the 10–50 µl region, relative standard deviations of 0.1% were found with polyethylene pipets made in the laboratory (6). Their design is shown in Fig. 6.

The pipets are calibrated with water. The pipet volume is determined by taking the mean of at least five weighings on a microbalance of the pipet contents and considering the density of water at that temperature. Small ground-glass stoppered flasks ($\bar{\mp}$ 5, cf. Fig. 7) are used for weighing vessels; their capillary necks (0.6–0.8 mm diameter) are closed with polyethylene stoppers to preclude evaporation of the water during the weighing.

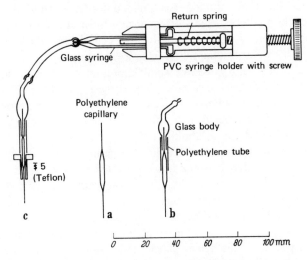

Figure 6. Polyethylene ultramicropipet for solution partition. The pipet body (*a*) is made by drawing out a piece of polyethylene tubing (1.8 mm o.d.) over the pilot flame of a Bunsen burner. The tubing is heated to the softening point while being rotated, removed from the flame, and drawn out slowly; the capillary diameter depends on the speed with which it is drawn out. (Seasoned polyethylene can be worked better than new.) The pipet is inserted into the hole of a Teflon stopper ($\bar{\mp}$5) and is connected to a 100 µl glass syringe via a piece of polyethylene tubing. By moving the piston with the screw, the pipet body can be filled or emptied.

1cm

Figure 7. Weighing vessel for pipet
calibration.

In the measurement of organic solvents, small, empirical corrections must be applied for their different wettabilities; their greater thermal coefficients of expansion must also be considered.

Polyethylene is not detectably attacked during the measuring process by water, methanol, or ethanol; acetone and ethyl acetate, on the other hand, dissolve small amounts which are, however, reproducible upon observance of equal measuring times. Thus, for example, required blanks in a carbon or hydrogen determination can readily be corrected after solution partition (cf. Sect. 1.2.4.4) by knowledge of the solubility.

For the measurement of aromatic or saturated hydrocarbons, as well as chlorinated hydrocarbons, polyethylene pipets are unsuitable, since these solvents dissolve considerable amounts of polyethylene. Glass pipets must then be used, which have a somewhat greater relative standard deviation of about 0.3%.

For measuring out aqueous or alcoholic reagent solutions, automatically filled polyethylene pipets with reservoirs (30)[3] are very suitable (cf. Fig. 8). Pipets are available for volumes between 0.5 and 250 μl. Volumes of solution greater than 20 μl can be measured with an error of less than 0.2%.

Grunbaum reports a self-filling type of glass pipet with reagent reservoir (32,33) (cf. Fig. 5, no. 8), which is likewise quick and easy to manipulate; with it there is also scarcely any danger of introduction of impurities.

For the transfer of solutions from one container to another, a rubber bulb pipet with a finely drawn tip (cf. Fig. 9) is used or the

---

[3] Manufacturer: Beckman Instruments.

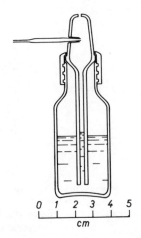

0 1 2 3 4 5
cm

Figure 8. Polyethylene pipet after Sanz. After squeezing the polyethylene bottle, the opening in the upper, bell-shaped part is closed with a finger and the solution drawn up through the capillary until it issues as a small drop from the outer end of the capillary and overflows. At this moment the opening is released and the pressure stopped. To pipet the liquid out, the upper opening is closed with the index finger and the liquid is then slowly ejected by squeezing the polyethylene bottle.

device shown in Fig. 10, which is particularly suited for transfer without loss of organic solvents into cuvets (34). A 1- or 2-ml transfer pipet with a finely drawn tip is fused at its other end to a 5-ml piston syringe which is bent downward. The pipet is fastened to a stand by a buret clamp.

1.2.4.2. BURETS

For measuring out standard solutions and for titration it is desirable to use micro piston burets with micrometer screws, dial gauges, or geared motors.

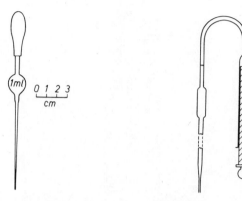

Figure 9. Rubber bulb pipet. **Figure 10.** Device for transferring organic solvents into cuvets.

For measuring solutions in the 50–500 μl region, accurately calibrated syringes (total volume: 100 or 500 μl) with closely ground glass pistons are suitable. Depending on the quality of the grinding, they have standard deviations of between 0.1 and 0.05 μl. Figure 11 shows such a glass piston buret, which is also available in other designs.[4]

Bent glass canulas can be attached to the ground joint of the syringe; these must dip into the test solution for continuous addition of titrant. Organic solvents can be measured accurately with these glass piston burets only with an added polyethylene or Teflon packing between the syringe body and the piston (cf. Fig. 12), for otherwise the solvents creep back in the ground glass and evaporate at the end.

For titration with solutions $>0.01N$ the open titration system shown in Fig. 13 is recommended (35). The titration vessel stands on a white base on a magnetic stirrer. For better recognition of the endpoint a daylight lamp should be used.

A variant[5] (34) is used in systems sealed from the laboratory air for titration with very dilute titrant solutions which are sensitive to oxygen, carbon dioxide, and water (cf. Fig. 14).

A Duranglas-50 or Pyrex syringe body (capacity: 500 μl) is

[4] For example, Agla Micrometer Syringe Outfit.
[5] Manufacturer: Glastechn. Werkstätte, Mainz, Rheinalle 28, Germany.

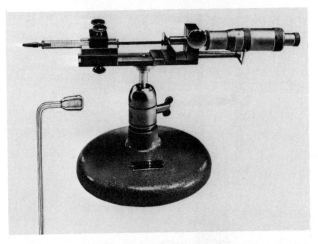

Figure 11. 100-μl glass piston buret of Braun Apparatebau, Melsungen, Germany.

fused to a three-way stopcock with a Teflon plug. The stopcock is connected to a reservoir (capacity: 75–100 ml). The ground opening of the reservoir can be closed with a drying tube. To the remaining leg of the capillary stopcock a thin polyethylene tube (diameter: 1.8 mm) is connected, through which the solution can flow into the titration vessel. For the buret tip, a finely drawn polyethylene capillary[6] is used which dips into the solution to be

---

[6] Polyethylene capillaries are prepared by softening 2-mm tubing with rotation a few centimeters over the pilot flame of a Bunsen burner and drawing it out slowly after removal from the heat. By varying the drawing rate it is possible, after some practice, to draw out capillaries of various diameters and bores.

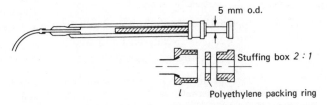

Figure 12. Glass syringe with stuffing box for measuring out organic solvents.

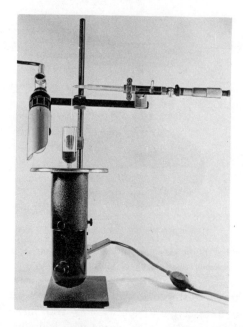

Figure 13. Titration system after Belcher.

titrated. The capillary must be drawn out so thin that the test solution cannot diffuse back perceptibly; on the other hand, no excess pressure should occur in the buret which might engender afterflow of the titrant.

Even 10^{-4} to $10^{-5}N$ titrant solutions can be kept at constant titer for a long time in carefully pretreated burets.[7] For light-sensitive solutions the glass parts should be lacquered black. Every glass part of a buret should always be used for solutions of the same composition and concentration; before use it should be conditioned for at least 12 hr with the titrant solution so that adsorption–desorption equilibrium can occur between walls and solution.

For automatic titration, the micrometer screw is replaced with

---

[7] The buret and reservoir are treated several times with concentrated hydrochloric and nitric acids alternately and subsequently with ammonia water. They are then left standing for several hours filled with double-distilled water; the water should be renewed several times and the buret then steamed for at least 30 min.

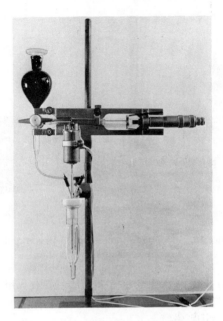

Figure 14. Titration apparatus with 500-μl glass piston buret with reservoir and micrometer screw (34).

a gear rack which is driven by a synchronous motor through multistage gearing[8] and can drive the buret piston at various rates. The piston advance can be coupled to the paper transport of a recorder or it can be calibrated by means of a revolution counter attached to the gearing (cf. Fig. 15).[5]

Aqueous and organic liquids of under 100 μl can be measured with the Sanz buret (30)[3] (cf. Fig. 16) with standard deviations of ± 0.007 or 0.008 μl; its principle represents a substantial improvement over the hitherto usual 100-μl piston burets (36,37). The high accuracy of this buret is attained by means of a Teflon displacement piston of precise diameter which, passing through a Teflon packing, can be moved in the glass body containing the solution and thus displaces a volume of solution corresponding to the piston movement.

---

[8] For example, gearing of Halstrup Co., Kirchzarten, Schwarzwald, Germany.

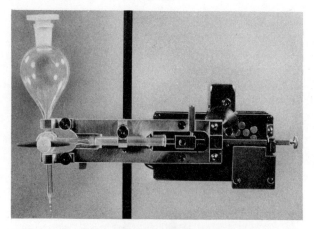

Figure 15. 500-μl piston buret with reservoir and ten-stage
gearing for automatic titration.

Grunbaum (38)[9] describes a similar buret which is provided
with a reservoir, analogous to the glass piston buret with reservoir
previously described.

Again, finely drawn polyethylene capillaries serve as buret tips.
For the purposes of organic elemental analysis these displacement
burets are especially valuable in measuring out small amounts of
standard solutions for calibration, since relatively concentrated and
thus more stable standard solutions can be used.

It is advantageous to employ little 10-ml two-necked flasks (cf.
Fig. 17) for filling burets without reservoirs. The polyethylene

---

[9] Manufacturer: Microchemical Specialties Co., Berkeley, California.

Figure 16. Ultramicroburet after Sanz.

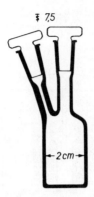

Figure 17. Flask for short-term storage of
standard solutions for calibration.

capillary of the buret is inserted through the narrow capillary neck
(0.8–1 mm i.d.) into the solution to be drawn up, during which the
solvent cannot perceptibly evaporate. The wider ground neck is for
filling the flask with a pipet. For measuring reagent solutions in
amounts of 0.5–2 ml, the usual microburets, with measuring capillaries and reservoirs with $\overline{\$}$ 14.5 openings, have proved good. Their
stopcocks must have Teflon plugs. For determinations that must
be carried out with laboratory air excluded, the reagents are con-
veyed to the titration vessel in thin polyethylene tubing.

A reservoir (cf. Fig. 18,c) provided with a magnetic valve (a)
and a discharge capillary (b) is suitable for automatic dispensing
of a reagent solution (>100 μl, reproducibility: 1–2%). To
measure out a specific amount of reagent, the magnetic valve is
opened for a precise length of time by exciting the electromagnet
(e) through an electromechanical or electronic time switch. The
time required to discharge a given amount of reagent is determined
beforehand. Change in the discharge rate caused by the varying
level of the reagent solution in the reservoir is substantially avoided
by applying the principle of the Mariotte flask. The diameter of the
discharge capillary depends on the required amount of reagent and
the accuracy of the time switch. In the control system described
in Sect. 2.2.2.3.6, a switching relay is actuated after an electromechanical counting register has counted a predetermined number
of pulses from a pulse generator (pulse rate: 100/min). With this

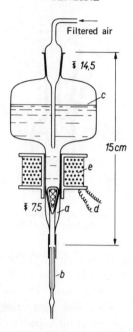

Figure 18. Dispensing buret with
magnetic valve.

system, a discharge time of at least 2 min is appropriate for measuring out an aqueous solution of 100–1000 μl to about 1%; the capillary diameter must be chosen accordingly.

In principle, other automatic solution-dispensing systems can be used, e.g., motor-driven piston burets or oxyhydrogen coulometer pipets (39–42). The former are, however, more complicated and the latter more strongly temperature dependent.

1.2.4.3. VOLUMETRIC FLASKS

For making solutions in a small, definite volume, little volumetric flasks of various designs are used (25), which are described in their specific sections.

Whenever possible, no volumetric flasks smaller than 5 ml should be employed, for they are troublesome to use and inaccurate. In the case of a solid that is to be dissolved in a specific volume of solvent, a definite amount of the solvent is added from a piston buret. For dilute solutions the change of volume upon solution may be neglected. When a decomposition product is to be brought to a certain solution volume, it is often possible with nonvolatile products to take to dryness the solution resulting from the decomposition and to take up completely the residue in an amount of solvent measured out from a buret (cf. Sect. 2.6.3.3.4).

Quartz decomposition tubes can be calibrated with a mark, thus permitting dilution to volume in the decomposition vessel itself (cf. Sect. 2.10.3.1.4).

In photometric determination procedures it is often possible to extract the colored compound from the aqueous phase into a known volume, measured with a buret, of an organic solvent and to examine the organic phase photometrically (cf. Sects. 2.6.3.3.4 and 2.10.3.1.4).

1.2.4.4. SOLUTION PARTITION (cf. Sect. 1.2.1.2)

The aluminum foil weighing boat (tare weight about 1 mg, cf. Fig. 4*b*) bearing the weighed sample is transferred from the weighing chamber (Fig. 3, left) into the preparation chamber (Fig. 3, center), which contains the apparatus for solution partition. The sample and foil are placed in the 1-ml two-necked $\overline{\mathrm{S}}$ flask (*1*) of the partition apparatus (Fig. 19). With a 500-μl micro piston buret, [10] say 200 $\pm$0.1 μl solvent (water, ethanol, acetone, *inter alia*) is added. After the sample has been dissolved, with stirring,[11] in the tightly closed (Teflon stopper) flask,[12] a definite fraction of the solution, between 10 and 40 μl, is removed with the micropipet [13] (Fig. 20, right) and transferred to a support, e.g., into a thin-walled quartz combustion vessel (cf. Fig. 4*c*), onto a paper disk (cf. Sect. 1.3.1.), onto quartz wool, into a platinum boat, or into a little decomposition flask (cf. Sect. 2.10.3.1). It is advisable to have the support on a holder with a soft-iron core, so that it may be held fast by a permanent magnet fastened to a stand. Thus the

---

[10] For measuring organic solvents, the buret must be provided with a polyethylene packing (cf. Fig. 12).

[11] For this purpose there is a small motor and magnetic stirrer in the base under the flask and a stirring bar in the flask.

[12] The volume change upon solution can be neglected.

[13] To minimize evaporation in the case of volatile solvents, the flask should be opened only briefly for the withdrawal.

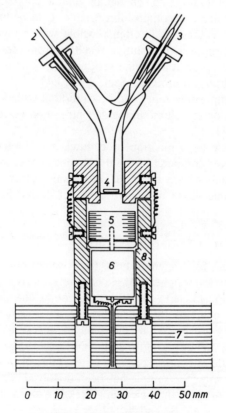

Figure 19. Solution partition
apparatus.

1. Flask
2. Teflon stopper with polyethylene
   tube to the piston buret
3. Teflon stopper with pipet
4. Stirring bar
5. Magnet
6. Motor
7. PVC base plate
8. PVC base

support can be positioned under the discharge capillary of the pipet
for transfer of the solution or under a fine polyethylene capillary to
blow away the solvent.

The stream of air used to blow away the solvent must be previ-

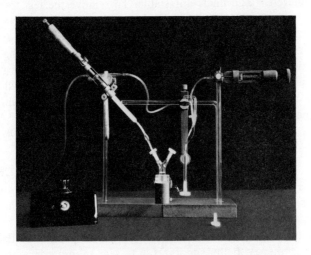

Figure 20. Overall view of the solution partition apparatus.

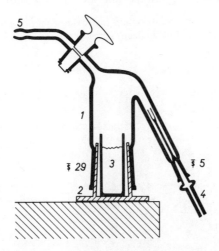

Figure 21. Quartz combustion vessel in microdesiccator: *1.* Microdesiccator. *2.* Polyethylene stopper. *3.* Vessel with P_2O_5. *4.* Quartz combustion vessel with holder. *5.* Connection for vacuum pump.

ously filtered [14] and finely adjustable. To remove the last traces of solvent (cf. C and H determination), the support bearing the sample is put into a small desiccator (Fig. 21) and the solvent drawn off with a vacuum pump or removed by freeze drying. The freeze-drying apparatus (Fig. 22) consists of a two-stage rotary vane pump after Gaede, with a vacuum gauge, and two cold-traps (liquid nitrogen) to which the microdesiccators can be connected via a manifold with 4–5 stopcocks (cf. Fig. 22).

1.2.5. TRANSFER OF LARGE VOLUMES OF SOLUTION ONTO SMALL SUPPORTS

Often solutions accumulate upon enrichment and separation operations, or the sample is dissolved in an inert, volatile solvent, e.g., from a cold finger, from chromatography paper (43), or from a thin-layer chromatogram. In some cases the sample can be decomposed together with the support (44). The solutions are concentrated to a few milliliters by thin-layer evaporators or other devices (45). The solution concentrates can be put onto the supports, paper disks, quartz wool, *inter alia,* in a definite sequence of drops

---

[14] A U-tube filled with solid sodium hydroxide connected to another filled with ignited quartz wool (50 cm deep) serves as a filter.

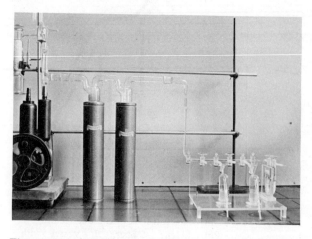

Figure 22. Freeze-drying system for solvent removal after solution partition.

by, e.g., piston burets of which the pistons can be moved slowly with the aid of geared motors. The solvent is blown away after each drop delivery. These dispensing devices have dead volumes which cannot be neglected, so they must be rinsed several times with pure solvent.

The following simple arrangement (46) (cf. Fig. 23) permits transfer of a solution onto a small support without time-consuming rinsing: The solution is forced out of a 6-ml $\overline{\mathbb{S}}$ vessel, through a capillary tube (c) (1.1 or 0.5 mm i.d.), and dropwise through a finely drawn polyethylene exit capillary by oxyhydrogen generated by electrolysis of a $0.1N$ potassium hydroxide solution in a connected cell (d). At constant temperature a uniform sequence of drops can be produced with the help of the electrolysis current. The exit capillary is in a glass evaporation chamber (a) over the support, which is brought into the chamber from the side. Under the support is a nozzle from which issues a filtered stream of air that blows the solvent away. To hasten the evaporation, the evaporation chamber can be warmed with the aid of an aluminum block (b).

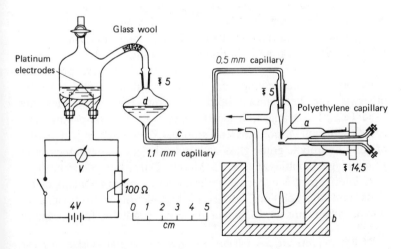

Figure 23. Arrangement for transfer of solutions onto small paper supports.

1.2.6. MANIPULATION OF VERY DILUTE SOLUTIONS

Several authors (2,21,25,47–52) refer to the difficulties of preparing and preserving, without change in concentration, very dilute ($<10^{-3}M$) solutions of exact concentration (standard and titrant solutions).

The principal causes of concentration changes, such as chemical decomposition, ion exchange with the container material, and wall adsorption, vary greatly from case to case and depend above all on the species of ion or molecule, the solvent, the acidity of the solution, the container material and its pretreatment, the temperature, and photosensitivity. Thus there are quite varied conditions for keeping solutions, which must be heeded exactly in the instructions.

Let us present here some generally valid rules:

The more concentrated the solutions used, the smaller the errors. It is always advantageous to measure out less dilute solutions in very small amounts (ultramicroburets or-pipets) rather than very dilute solutions with ordinary burets or pipets.

The solutions should be prepared as more concentrated stock solutions and brought to the requisite concentration with calibrated quartz pipets and volumetric flasks just before use. Only thoroughly cleaned and steamed flasks should be used, which have been shaken several hours with solutions of the same final concentration. If volumetric flasks are used which have not been brought into equilibrium with the solution, then immediately after filling to the mark and thorough mixing the solutions must be transferred to storage bottles, the walls of which have previously been brought into equilibrium with the dilute solution. Unless otherwise specified, storage bottles of brown borosilicate glass are used (e.g., Fa. Jeanaer Glaswerke, Mainz), with $\overline{\$}$ openings that are closed with $\overline{\$}$ polyethylene stoppers,[15] or else reagent bottles of good hydrolytic grade glass with polyethylene screw-cap closures (e.g., reagent bottles of Fa. E. Merck, Darmstadt, Germany).

If interference is expected by introduction of components of the glass into the solution, bottles of high-density polyethylene or polypropylene are used. Neither material is suitable for storage of

---

[15] Glass stoppers are not suitable, for they do not seal satisfactorily and cause glass abrasion.

dilute solutions of oxidizing agents, e.g., OBr^-, MnO_4^-, Ce^{4+}, CrO_4^{2-}. On the other hand, 10^{-4} to $10^{-5}M$ acetic acid solutions of $AgNO_3$ and neutral KCl, KBr, and KI solutions can be kept at constant titer for several weeks in polyethylene bottles that have been previously conditioned with the solutions and painted black for protection from light. In some cases quartz bottles are necessary.

New bottles should be treated several times before use alternately with concentrated hydrochloric and nitric acids and subsequently with ammonia water. They are then left filled for several hours with double-distilled water that is renewed from time to time. The bottles should then be steamed for at least 30 min. Solutions are normally kept in dust-proof boxes (cf. Fig. 3, center); solutions sensitive to heat, e.g., sodium hypobromite, are kept in a refrigerator.

For preparation of aqueous solutions it is best to use double- or triple-distilled water from a quartz still (e.g., from Fa. Heraeus, Hanau, Germany). Single-distilled water which has subsequently been demineralized with an ion exchanger is not suitable in most cases, for it contains too large amounts of organic matter.

Organic solvents must be freshly distilled before use through an effective column; in certain cases additives are employed before the distillation for the reduction of ionic impurities (e.g., $AgNO_3$ for retention of halides).

It is a basic principle that reagent, standard and titrant solutions should be taken from their storage vessels only by means of clean, previously conditioned pipets; in doing so, the bottles should be removed from their dustproof boxes only briefly.

1.2.7. REFERENCES

1. A. A. Benedetti-Pichler, *Handbuch der Mikrochemischen Methoden*, Vol. 1, Part 2, Springer, Wien, 1959.
2. I. M. Korenman, *Introduction to Quantitative Ultramicroanalysis*, Academic Press, New York, 1965.
3. S. L. Bonting and B. R. Mayron, *Microchem. J.,* **5**, 31 (1961).
4. A. M. Burt, *Microchem. J.,* **11**, 18 (1966).
5. H. M. El-Badry, *Micromanipulators and Micromanipulation*, Springer, Wien, 1963.
6. K. H. Ballschmiter and G. Tölg, *Z. Anal. Chem.,* **203**, 20 (1964).

7. P. L. Kirk, *Quantitative Ultramicroanalysis*, 2nd ed., Wiley, New York, 1951.
8. R. C. Wrigley and C. E. Gleit, *Anal. Chem.*, **36**, 307 (1964).
9. G. Gorbach, *Mikrochemie (Wien)*, **14**, 254 (1936).
10. G. Gorbach, *Microchem. J.*, **10**, 485 (1966).
11. K. Behrndt, *Z. Angew. Physik*, **8**, 453 (1956).
12. G. Ingram, *Ind. Chemist*, **37**, 343 (1961).
13. H. V. Neher, in *Procedures of Experimental Physics*, J. Strong, Ed., 2nd ed., Prentice-Hall, New York, 1942.
14. P. L. Kirk, R. Craig, J. E. Gullberg, and R. Q. Boyer, *Anal. Chem.*, **19**, 427 (1947).
15. H. M. El-Badry and C. L. Wilson, *Roy. Inst. Chem. (London)* Report No. 4, **1950**, 23.
16. J. A. Kuck, P. A. Altieri, and A. K. Towne, *Mikrochim. Acta*, **1953**, 254.
17. I. M., Korenman, Y. N. Fertel'meyster, and A. P. Rostokin, *Zavodsk. Lab.*, **16**, 800 (1950).
18. H. Carmichael, *Can., J. Phys.*, **30**, 524 (1952).
19. R. Bowers and E. A. Long, *Rev. Sci. Instr.*, **26**, 337 (1955).
20. H. Asbury, R. Belcher, and T. S. West, *Mikrochim. Acta*, **1956**, 598.
21. I. P. Alimarin and M. N. Petrikova, *Anorganische Ultramikroanalyse*, Deutscher Verlag der Wissenschaften, Berlin, 1962.
22. R. Belcher, *Submicro Methods of Organic Analysis*, Elsevier, Amsterdam, 1966.
23. T. R. F. W. Fennell and J. R. Webb, in *Microchemical Techniques*, N. Cheronis, Ed., Vol. 2, Wiley, New York, 1962, p. 1003.
24. W. J. Kirsten, *Z. Anal. Chem.*, **181**, 1 (1961).
25. J. Mika, *Die Methoden der Mikromassanalyse, Die Chem. Analyse*, Vol. 42, Ferd. Enke-Verlag, Stuttgart, 1958.
26. D. Glick, *Techniques of Histo-and Cytochemistry*, Interscience, New York, 1949, p. 170.
27. H. Mattenheimer, *Mikromethoden für das klinisch-chemische und biochemische Laboratorium*, 2nd ed., de Gruyter, Berlin, 1966.
28. D. J. Prager, R. L. Bowman, and G. G. Vurek, *Science*, **147**, 606 (1965).
29. H. Mattenheimer and K. Borner, *Mikrochim. Acta*, **1959**, 916.
30. M. C. Sanz, *Chimia (Aarau)*, **13**, 192 (1959).
31. R. C. Sisco, B. B. Cuningham, and P. L. Kirk, *J. Biol. Chem.*, **139**, 1 (1941).
32. B. W. Grunbaum and P. L. Kirk, *Anal. Chem.*, **27**, 333 (1955).

33. B. W. Grunbaum, *Microchem. J.*, **9**, 46 (1965).
34. B. Morsches and G. Tölg, *Z. Anal. Chem.*, **200**, 20 (1964).
35. R. Belcher, *Submicro Methods of Organic Analysis,* Elsevier, Amsterdam, 1966.
36. P. F. Scholander, *Science,* **95**, 177 (1942).
37. R. Gilmont, *Anal. Chem.*, **25**, 1135 (1953).
38. B. W. Grunbaum, *Microchem. J.*, **9**, 245 (1965).
39. G. C. Ware, *Lab. Prac.*, **6**, 656 (1957).
40. R. J. Heckly, *Science,* **127**, 233 (1958).
41. W. Mathias, *J. Chromatography,* **3**, 501 (1960).
42. K. Beyermann, *Z. Anal. Chem.*, **210**, 1 (1965).
43. A. F. Krivis, G. E. Bonson, W. A. Struck, and J. L. Johnson, *Microchem. J., Symp. Ser.*, **1**, 293 (1960).
44. H. Soep, *J. Chromatog.*, **6**, 122 (1961).
45. H. Frehse and H. Niessen, *Z. Anal. Chem.*, **192**, 94 (1963).
46. G. Tölg, *Z. Anal. Chem.*, **194**, 20 (1963).
47. G. G. Eichholz, A. E. Nagel, and R. B. Hughes, *Anal. Chem.*, **37**, 863 (1965).
48. G. H. Morrison, *Trace Analysis,* Interscience, New York, 1965.
49. J. H. Yoe and H. J. Koch, *Trace Analysis,* Wiley, New York, 1957.
50. O. G. Koch and G. A. Koch-Dedic, *Handbuch der Spurenanalyse,* Springer, Berlin, 1964.
51. I. M. Korenman, *Analytical Chemistry of Low Concentrations.* Engl. Transl. from the Russian by Schmorak; Vol. 7. Israel Program for Scientific Translations, Jerusalem, 1969.
52. G. Tölg and I. Lorenz, "Methoden der mikrochemischen Elementbestimmung und ihre Grenzen," in *Fortschritte der chemischen Forschung,* Vol. 11/4, Springer-Verlag, Berlin-Heidelberg-New York, 1969.

1.3. DECOMPOSITION PROCEDURES, GENERAL

With decreasing sample size, the danger rapidly increases of losing uncontrollable amounts of the element to be determined (e.g., by adsorption on the vessel walls, ion exchange, and volatility) and introducing impurities; this is already true during decomposition of the sample.

Especially in decomposition of samples of less than 30 μg, difficulties are to be expected which, caused primarily by introduced contaminants, increase from element to element in approximately the order: I, Br, F, P, S, Cl, N, C, H, O.

Easily purified gases, such as oxygen or hydrogen, are always preferable to solid or liquid decomposition reagents which, in the usual "analytical reagent" grade, may contribute considerably to blank values (cf. Table 4). In consideration of these blank values, no more solid or liquid reagent should be used in decompositions than the amount absolutely necessary.

Just as important as purification of reagents is proper storage in resistant vessels with effective dust protection (2–4,17). It is a basic principle that double-distilled water from a quartz still should be used for preparation of aqueous solutions.

Similar requirements apply to the decomposition apparatus. Next to the proper choice of materials, the minimization of surface areas and the cleanliness of equipment are of crucial importance. Quartz apparatus is generally preferable to glass. Teflon and, in certain cases, polyethylene are good for stoppers. Ground-glass

Table 4. Maximum Impurity Content of Common Decomposition Reagents (μg/g) (1)

	SO_4^{2-}	Cl^-	PO_4^{3-}	Total N
NaOH a.r.	5	5	5	3
Na_2O_2 a.r.	10	100	5	30
H_2O_2, 30% a.r.	2	0.8	3	2
H_2SO_4, 96% a.r.		0.5		0.2(NO_3^-)
H_2SO_4, 96% ultrapure			0.01	
$HClO_4$, 70% a.r.	10	3	50	40
$HClO_4$, 70% ultrapure		0.5	0.1	
HNO_3, 65% a.r.	1	0.5		
HNO_3, 65% ultrapure	0.5	0.5	0.01	
Sodium a.r.	20	20	10	5

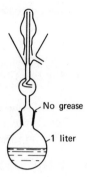

Figure 24. Apparatus for steaming
vessels.

joints with organic sealants (e.g., Apiezon or silicone grease)
should be avoided. Only polyethylene, polypropylene, or Teflon is
suitable for tubing.

It is usually adequate to clean the apparatus by boiling in $6N$
hydrochloric acid or, if chloride ion might interfere, in concentrated
nitric acid, then to rinse it thoroughly with distilled water followed
by double-distilled water, and to store it protected from dust.[16]
Before use, the equipment should be steamed again (about 30 min),
using an all-glass steam generator (cf. Fig. 24) and distilled water.
Finally, in order that as few steps as possible be required for the
execution of a procedure, one should endeavor to obtain the element from the decomposition in the ultimate determination form
and to carry out the decomposition and determination in the same
apparatus.

Very small amounts of C, H, O, N, Cl, Br, and S can be determined with sufficient accuracy only if the decomposition and
determination are conducted in apparatus that is completely sealed
against the laboratory air and if the samples are burned in purest
oxygen or hydrogenated in purest hydrogen.

Decompositions for the determination of the remaining elements
are subject to less interference by introduced contaminants. In
determinations of less than 1 μg P, F, and I, the samples can be
decomposed in the particularly simple oxygen flask (5–7) or (e.g.,
for P determination) by perchloric acid-nitric acid or perchloric

---

[16] The treatment of vessels with chromosulfuric acid is categorically in-
advisable, for considerable amounts of chromium are adsorbed by the
walls and cannot be removed even by assiduous rinsing with water.

acid–sulfuric acid in an open decomposition vessel if all manipulations are conducted with dust excluded.

The use of the oxygen flask is also possible in determinations of S, Cl, and Br content in samples of over 30 μg; thus this method is then applicable to the determination of S, F, Cl, Br, I, and P.

In order to avoid repetition of the description of often-used decomposition procedures in each specific chapter, the following sections will summarize the more important procedures.

1.3.1. DECOMPOSITION IN THE OXYGEN FLASK

Belcher and co-workers (8) scaled down the flask volume for decomposition of milligram amounts to the dimensions suitable to samples of over 30 μg (cf. Fig. 25). Instead of wrapping the sample in paper strips, thin polyethylene sheet (8) or the collodion film "sandwich technique" (18) is used to achieve the smallest possible blank values. A linen thread serves to ignite the polyethylene sample packet, which is fastened to a platinum wire holder in the flask containing oxygen and absorbent solution; the thread is lit by a pure alcohol flame before introduction of the sample into the flask.

A platinum wire (50 mm × 0.3 mm) is fused to the $\overline{\mathfrak{S}}$ 14.5 hollow glass stopper of the oxygen flask (Fig. 25). At the other

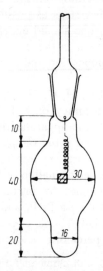

Figure 25. Oxygen flask after Belcher (dimensions in millimeters).

end of the wire is a piece of platinum gauze (80 mesh, 13 × 3.5 mm), held in place by the wire wrapped around it; it should not be fastened by spot welding, which might cause losses (e.g., in chloride determination). The gauze should be in the center of the wide portion of the flask. Before use the flask should be treated for 5–10 min with 2% hydrofluoric acid, thoroughly washed with hot water, and steamed for at least 10 min. The apparatus shown in Fig. 24 can be used for this procedure.

The ground stopper with platinum wire and gauze should be thoroughly rinsed with distilled water and purest alcohol, and the platinum parts vigorously ignited with a Bunsen flame. The combustion flask should immediately be closed with the stopper and kept in an empty desiccator until use. It is advisable to repeat this procedure after every determination.

Polyethylene sheet. To prepare very thin polyethylene sheet (about 0.01 mm), a piece of tubing, previously thoroughly cleaned with water and alcohol, is carefully heated over a flame and blown up to a large balloon. This is cut into 15-mm squares, each weighing about 2 mg. The pieces are stored between filter paper for protection from dust and chloride.

Linen thread. A linen thread wound around a microscope slide is soaked overnight in a 1:1 ethanol–water mixture, thoroughly rinsed with this mixture, and dried at 110° for 3 hr. The thread should be stored in a tightly closed bottle and cut to length only before use.

The Kirsten technique (9) simplifies this procedure and moreover leads to smaller blank value variations. Kirsten obviates the combustible sample support (polyethylene sheet or collodion film) and the ignition of the sample by bringing the sample in a little quartz combustion tube into a horizontal combustion vessel of Supremax or, preferably, quartz tubing which has been heated by a tube furnace to 850° (cf. Fig. 26). The absorbent solution is in a cold zone of the combustion vessel, outside the furnace. This design, which is described for the decimilligram region, has already been successfully tested in the centimilligram region (10,11).

For the determination of F, I, S, and P in amounts under 1 μg, the following procedure (12,13) is given: The combustion flask is a 10-ml quartz vessel with $\overline{\text{\$}}$ 14.5 opening (cf. Fig. 27a), narrowed at the lower end so that it may be inserted into the hole of a

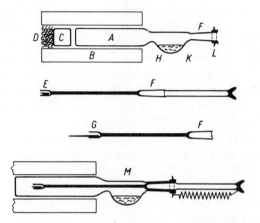

Figure 26. Apparatus for decomposition in the oxygen flask after Kirsten (9). (*A*) Quartz decomposition vessel, about 14 mm i.d., about 1.5 mm wall, about 60 mm projecting into the furnace; (*B*) tube furnace, at 850°; (*C*) quartz spacer; (*D*) asbestos: (*E*) quartz sample holder with $\overline{\overline{\mathrm{S}}}$ 7.5 joint (*F*); (*G*) sample holder with Supremax glass capillary for liquid samples; (*H*) absorbent solution in depression (*K*); (*L*) metal band with hooks for fastening springs; (*M*) the apparatus during decomposition.

heating block for heating and evaporating solutions. The dissolved sample, as described in Sects. 1.2.4.4 and 1.2.5, is transferred to a paper disk[17] of 5 mm diameter, which is clamped in the platinum loop of the electrode holder shown in Fig. 27*b*. Before tightly closing the flask with the Teflon stopper on the electrode holder, the absorbent solution is put in and the flask filled with pure oxygen. All operations should be conducted in a glovebox (cf. Fig. 3). The

---

[17] Paper disks of 5 mm diameter can be punched out of quantitative ashless filter paper (e.g., Schleicher and Schüll, White Band No. 589) with a cork borer or paper punch and are then treated a few minutes with 6N hydrochloric acid, which should be renewed periodically. Finally, the disks are washed free of chlorides with double-distilled water in the absence of dust; the water should be renewed about ten times in the course of several hours. The disks are dried at 110° and stored protected from dust. They should then be touched only with forceps.

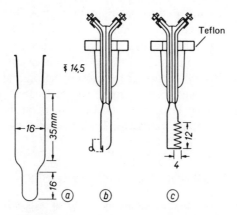

Figure 27. Combustion apparatus with electrical ignition (13, 14) (dimensions in millimeters).

amount and kind of absorbent solution to be used depend on the element to be determined.

The paper is ignited by means of an induction coil (about 10 kV, e.g., an automobile ignition coil with interrupter). To absorb the combustion products, the flask should be shaken in a shaking machine or left standing for at least 30 min, during which the walls should occasionally be wet with the absorbent solution by inclining and rotating the flask. In the case of difficulty combustible substances, e.g., sulfonic acids and their salts, paper disks treated with picric acid effect higher combustion temperatures and thus more complete combustion.[18]

Smaller blank values in decomposition are obtained with platinum foil (13) or cleaned quartz wool (14) as support material than with paper or polyethylene.[19] Very small blank values have been attained, especially with quartz wool. The quartz wool is inserted into a spiral (4 mm i.d. $\times$ 12 mm length) wound of about 30 cm of 0.3 mm Pt–Rh wire. Current is supplied to the spiral through two platinum leads (0.5 mm diameter) which are fused

---

[18] The disks are saturated with a 15% solution of picric acid in methanol and dried at 80–100°.

[19] Paper blank values: for I, Br, F, and P < 0.01 μg/0.2 cm²; for S about 0.05 μg/0.2 cm²; for Cl about 0.5 μg/0.2 cm² (without HCl treatment).

into a glass tube and passed gas-tight through a Teflon stopper (cf. Fig. 27c). To ignite the sample the spiral is heated as rapidly as possible to over 1000° with a low-voltage current source (10–20 V). The power consumption of the spiral is about 50 W. The evaporating sample must contact the glowing spiral and thus burns completely. The incandescence should last at least 15 sec, during which the combustion flask is cooled by immersion in cold water. With this method, blank values for I, F, P, and S were all under 0.005 μg.

1.3.2. DECOMPOSITION IN FLOWING GASES

In order to avoid blank values as much as possible in the determination of small amounts of C, H, O, N, Cl, Br, and S, the decomposition of the samples in purest oxygen or hydrogen is suggested (14,15), modeled on a micromethod of Ingram (16) for C and H determination.

The sample is decomposed within a Pt–Rh spiral, electrically heated to over 1000°, inside the wide portion of a quartz decomposition tube. The oxygen or hydrogen volume of about 3 ml is ample for the instantaneous reaction of the sample. Subsequently the reaction product is conducted by means of a purging gas, directly or after further reaction, to an absorption system and is there determined.

The gas stream (O_2, N_2, or H_2) is divided upon entering the apparatus; it passes through the upper and lower parts of the vessel simultaneously (cf. Fig. 28a). In the center of the widened chamber is the Pt–Rh heating coil (3) on a holder (2), which can be introduced through the ground opening of the decomposition vessel. The heating coil is wound of 40 cm of 0.3-mm Pt–Rh wire; it has an i.d. of about 4.7 mm and a length of about 15 mm. Current leads of 0.5-mm platinum wire are welded to each end of the spiral. The longer platinum lead is insulated from the spiral by a piece of quartz tubing. Current is introduced through the holder from a variable autotransformer (0–24 V, 200 W).

The sample is contained in a small quartz combustion vessel (5) on a holder. Before the decomposition it is inserted through the lower opening of the decomposition vessel into the cold coil, with the stopper of the holder then closing the opening. Only two-thirds

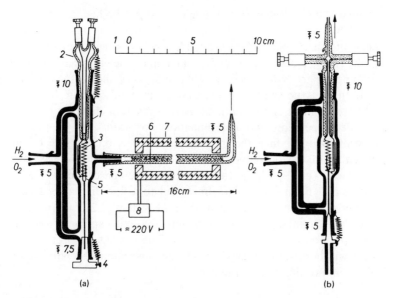

Figure 28. Apparatus for decomposition in flowing gases. *a*, with contact tube; *1*, quartz decomposition vessel; *2*, refractory glass holder with current leads to the platinum-rhodium spiral (*3*); *4*, Teflon stopper with holder for the quartz combustion vessel (*5*); *6*, contact tube; *7*, furnace; *8*, temperature controller; *b*, corresponding apparatus without contact tube.

of the quartz combustion vessel should be within the spiral.[20] For the decomposition the coil is brought in a fraction of a second to the requisite decomposition temperature, using a previously established emf versus temperature curve for oxygen or hydrogen. A spiral will, as a rule, last for over 300 decompositions; burned-out spirals can be quickly replaced.

There are two alternatives for the gas discharge from the system, depending on the nature of the element to be determined:

If, as in the case of Cl and Br determination, the reaction products (HCl or HBr) are highly water soluble and would thus enter into solubility equilibrium with the water film on the discharge

---

[20] When the current is turned on, heat conduction causes a temperature gradient in the quart vessel in the spiral; thus the sample in the lower part of the combusion vessel can never evaporate from the ignition zone of the spiral before burning.

capillary, then discharge is through the spiral holder, as shown in Fig. 28b. When the coil is glowing, the platinum leads conduct heat into the upper part of the holder and thus heat the entire gas path. This is aided by a silver wire (1 mm diameter) wound around the upper part of the decomposition tube (not shown for clarity).

If there is no danger of reaction product adsorption (e.g., for CO_2) or if there is an additional heated reaction chamber between the decomposition and determination apparatus (H, N, and S determinations), then the gas exit capillary is on the side of the widening of the decomposition vessel opposite the two gas inlet capillaries (cf. Fig. 28a). In both designs the end of the exit capillary is a $\overline{\math{S}}$ 5 capillary joint, drawn out to a fine discharge nozzle. The nozzle must be so fine that, with a gas flow of 2–3 ml/min, no absorbent solution can flow back into the exit capillary from the absorption vessel placed on the joint. The various absorption systems will be described later, as will the generation and purification of decomposition and purging gases.

1.3.3. EVAPORATION OF DIGESTS

The following evaporation apparatus[21] (Fig. 29) is recommended for heating or concentrating solutions at a specific temperature in the absence of dust and air-borne impurities or for evaporating decomposition acids before further treatment.

A pure aluminum or acid-resistant stainless steel heating block (a) of the size and shape shown in Fig. 29 is inside a heat-resistant vacuum desiccator (2.5-liter capacity). The block is electrically heated by an 80-W heating cartridge (b) inserted into a horizontal hole in its lower half. To prevent interference from evaporating metals, the heater is fused into a quartz tube (c). Current is supplied through the narrow part of the quartz tube which, sealed with a piece of polyethylene tubing, passes through a hole in the side of the desiccator.

By means of a relay and a contact thermometer (f), inserted into the heating block from above, any desired temperature to 300° may be obtained. To remove solvent vapors from the desiccator, an aspirator is connected at d which draws in purified air (cf. Fig. 3) or an inert gas, e.g., nitrogen, through opening e. It is

---

[21] Manufacturer: Glastechnische Werkstätte, Mainz, Rheinallee, Germany.

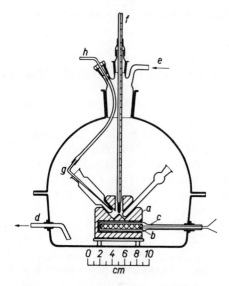

Figure 29. Apparatus for dust-free evaporation of solutions at specific temperatures. (*a*) aluminum heating block; (*b*) heating cartridge; (*c*) quartz tube into which the heating cartridge is fused; (*d*) aspirator connection; (*e*) air filter connection; (*f*) contact thermometer; (*g*) boiling capillary; (*h*) gas inlet.

furthermore possible to hasten evaporation and prevent boiling lags by bubbling a gas, e.g., nitrogen, into the solution through a quartz capillary (*g*). The gas is admitted at *h*.

1.3.4. REFERENCES

1. E. Merck, Darmstadt; Chemikalienkatalog 1966.
2. O. G. Koch and G. A. Koch-Dedic, *Handbuch der Spurenanalyse,* Springer, Berlin, 1964.
3. G. H. Morrison, *Trace Analysis,* Interscience, New York, 1965.
4. J. H. Yoe and H. J. Koch, *Trace Analysis,* Wiley, New York, 1957.
5. W. Hempel, *Angew. Chem.,* **13**, 393 (1892).
6. O. Mikl and J. Pech, *Chem. Listy,* **46**, 382 (1952); **47**, 904 (1953).

7. W. Schöniger, *Mikrochim. Acta,* **1955,** 123; **1956,** 869.
8. R. Belcher, *Submicro Methods of Organic Analysis,* Elsevier, Amsterdam, 1966.
9. W. J. Kirsten, *Microchem, J.,* **7,** 34 (1963).
10. T. R. F. W. Fennell and J. R. Webb, *Z. Anal. Chem.,* **205,** 90 (1964).
11. M. E. Fernandopulle and A. M. G. Macdonald, *Microchem. J.,* **11,** 41 (1966).
12. B. Morsches and G. Tölg, *Z. Anal. Chem.,* **200,** 20 (1964).
13. G. Tölg, *Z. Anal. Chem.,* **194,** 20 (1963).
14. G. Tölg, Habilitationsschrift, Mainz, 1965.
15. K. H. Ballschmiter and G. Tölg, *Z. Anal. Chem.,* **203,** 20 (1964).
16. G. Ingram, *Microchem. J. Symp. Ser.,* **2,** p. 495 (1962).
17. G. Tölg and I. Lorenz, "Methoden der mikrochemischen Elementbestimmung und ihre Grenzen," in *Fortschritte der Chemischen Forschung,* Vol. 11/4, Springer-Verlag, Berlin-Heidelberg-New York, 1969.
18. B. Griepink und W. Krijgsman, *Mikrochim. Acta,* **1968,** 330.

2. PROCEDURES FOR DETERMINATION OF THE ELEMENTS

2.1. GENERAL

The procedures for determination of the individual elements treated in the following sections are based on the original instructions, modified slightly in some cases when this seemed appropriate.

Comparative evaluations of the several procedures for determination of an element are hardly possible, since either the procedures require very different sample sizes or too few data exist for statistical error analyses.

Interferences and time required for a procedure will be dealt with to the extent that information is available.

For procedures of the author of this book, absolute standard deviations (1–4) are quoted, calculated for a finite number of determinations according to

$$s_n = \sqrt{\frac{\Sigma_i \ (x_i - \bar{x})^2}{n - 1}}$$

where x_i = each measured value; $\bar{x}$ = mean of all $x_i = (x_1 + x_2 + x_3 + \ldots + x_n)/n$; and n = the number of determinations. Thus, for example, s_{15} indicates that the calculation of the standard deviation is based on fifteen determinations carried out under uniform conditions.

It must be realized, however, that a statistical error analysis for ultramicroprocedures, which to some extent assume great manual dexterity and much practice, is problematic. In certain cases it is valid only for the one who worked out the procedure, while for the novice it is to be considered only an attainable value. Its statement is more meaningful, the fewer the manipulations subject to personal influence in a procedure. The same is true for statements of dispersion and confidence interval. The accuracy of a result (systematic error) in ultramicroprocedures, where the amount of sample available is always very small, is usually checked with known samples of similar composition; the theoretical value must agree with the determined value within the dispersion of the statistical error of the procedure. As far as possible, however, one should try to confirm a result by comparing it against at least one other obtained by an independent procedure. The value of knowing several procedures for the same element is thus evident; hence their comprehensive description is justified and the development of further procedures should always be welcomed.

2.1.1. REFERENCES

1. H. Kaiser and H. Specker, *Z. Anal. Chem.*, **149**, 46 (1956).
2. K. Doerffel, *Beurteilung von Analysenverfahren und -ergebnissen,* Springer, Berlin, J. F. Bergmann, München, 1962.
3. G. Gottschalk, *Einführung in die Grundlagen der chemischen Materialprüfung,* S. Hirzel, Stuttgart, 1966.
4. K. Doerffel, "Statistik in der analytischen Chemie," VEB Deutscher Verlag für Grundstoff-Industrie Leipzig, 1966.

2.2. CARBON AND HYDROGEN

2.2.1. DECOMPOSITION AND DETERMINATION, GENERAL

The elements are determined as carbon dioxide and water, obtained by combustion in pure oxygen. Combustion additives that release oxygen should be avoided as much as possible. The combustion apparatus (1) described in Sect. 1.3.2 leads to especially small blank values, while decompositions in sealed quartz tubes filled with oxygen, which proved good in the 100-μg region (2,3), or in a specially developed oxygen flask (4) lead to too large blank value variations.

The determination of very small amounts of carbon dioxide and water is carried out as follows:

1. Belcher and co-workers (5,6) avoid the difficulties of chemical determination by the employment of manometric procedures. Carbon dioxide is determined manometrically first (5) and then hydrogen (6), the latter in an apparatus of plain soda lime glass, which absorbs particularly little water, using Teflon-membrane valves in place of glass stopcocks. The measuring system is kept at a constant temperature of 50 ±0.1°. Water and carbon dioxide, formed in the sample decomposition in an oxygen stream at 1100°, are frozen out in cold traps at −80 and −196°. Sulfur and halogen compounds are fixed after combustion on a Körbl catalyst (7). Interferences by nitrogen oxides, arising from the combustion of nitrogen-containing samples, must be removed by an absorption system (chromosulfuric acid on a SILOCEL support (8) and manganese dioxide) between the carbon dioxide and water cold traps. After the remaining oxygen is pumped off, the coolants are removed and the gases allowed to expand into known volumes. Their pressures are measured with specially developed mercury manometers, in which the mercury required for pressure compensation is taken from piston burets. The indication of the mercury meniscus occurs photoelectrically (9). This method permits the determination of carbon and hydrogen with standard deviations of about ±0.08 μg C and ±0.05 μg H (calculated from ten values each, of Table 1 of the original work (6)). Adequate accuracies are attained with samples of about 50 μg. For the simultaneous determination of smallest amounts of carbon, hydrogen and nitrogen in routine analysis a gaschromatographic method (26) is of important interest.

2. Tölg and Ballschmiter (1,10) use samples of between 10 and 40 μg for the determination of carbon and of hydrogen. The standard deviations of these determinations amount to about ± 0.06 μg C and ± 0.01 μg H. Neither procedure is subject to interference by N, S, Cl, Br, I, or P.

Carbon determination (1). The carbon dioxide formed by combustion in pure oxygen is absorbed from the purging gas stream (10 ml/min) into a 0.01 or 0.005M barium hydroxide solution by finely dividing the gas in the absorbent solution with a bell-type stirrer (11). The absorption proceeds completely so long as the Ba^{2+}/CO_2 ratio is not materially less than 2:1. The solubility of the barium carbonate formed can be decreased by about two orders of magnitude by addition of water-soluble organic solvents such as 1,4-dioxane. The actual determination takes place by back-titration of either the remaining OH^- or the Ba^{2+} ions. The titration of Ba^{2+} ions with EDTA (disodium salt of ethylenediaminetetraacetic acid) (12) has proved superior. Even very dilute EDTA solutions have good titer constancy.

The titration takes place in water–dioxane mixtures. Dioxane, as compared to other water-soluble solvents, has a relatively slight volatility. It is added to the barium hydroxide solution after the CO_2 absorption as a prepared mixture with phthalein purple indicator and piperidine buffer. An advantageous composition for a sharp endpoint occurs when the solution to be titrated contains 60% dioxane, 3% piperidine, and 60 μg phthalein purple per milliliter upon addition of the mixture. The dioxane content may vary by $\pm 10\%$. To prevent a greater change in the dioxane concentration during the titration, 20% dioxane is added to the EDTA solution. (EDTA would precipitate at a higher dioxane concentration.) Slight differences of indicator concentration cause a displacement of the endpoint; the indicator solution must therefore be added with an accuracy of $\pm 1\%$ (1-ml microburet).

The colorless indicator solution (dioxane, water, piperidine, and phthalein purple) is stable for several weeks in diffuse daylight. The EDTA and barium hydroxide solutions, protected from atmospheric carbon dioxide by soda lime tubes, are titer constant for at least 4 weeks in the reservoirs of the piston burets.

Using 0.01 M solutions, photometric endpoint indication and carrying out the titration by hand, $s_{25} = \pm 0.04$ μg C (1-ml determination volume); using a motor-driven buret and a recorder, $s_{10} = \pm 0.02 – 0.03$ μg. Linearity between the quantity of CO_2 and the EDTA used for a 1-ml volume of solution to be titrated exists only for <20 μg C; above this the precipitated $BaCO_3$ increasingly influences the location of the endpoint.

Hydrogen determination (10). For the determination of about 10 μg water, corresponding to about 1 μg hydrogen, known procedures for larger amounts (13–21) have proved too insensitive or too inaccurate. Therefore a reaction was employed in which the product water is quantitatively converted to hydrogen sulfide by carbon disulfide vapor in a nitrogen stream at 500° on a corundum catalyst according to the following equation (22,23):

$$2H_2O + CS_2 \rightarrow 2H_2S + CO_2$$

The hydrogen sulfide can be accurately determined in even very small amounts. To remove the excess oxygen remaining from the decomposition, the water must first be frozen out at about $-170°$ and the apparatus flushed with pure nitrogen. The determination of water through hydrogen sulfide requires an exact determination method for about 5–25 μg hydrogen sulfide, from which the carbon disulfide must first be removed.

A complete CS_2 – H_2S separation is attained if the hydrogen sulfide is bound as zinc sulfide at pH 5–7 and the solution rinsed free of carbon disulfide by oxygen-free nitrogen.

The sulfide is subsequently titrated argentometrically (cf. Sect. 2.5.2.2.2.). For this the zinc absorbent solution (250 μl) is mixed with 250 μl of 0.25M EDTA, 0.5N in sodium hydroxide and 2N in ammonia, in which the precipitated zinc sulfide is soluble.

The total volume for determination amounts to 500 μl. Titration is with an acetic acid solution of 0.005 or 0.01N silver nitrate from a 500-μl micro piston buret with reservoir.

For the endpoint indication a differential potentiometric apparatus is used with two identical silver electrodes, polarized by a weak direct current (10^{-9} to 10^{-10} A $\cdot$ cm^{-2} current density). The method is explained in Sect. 2.7.2.2 and the circuit illustrated in Fig. 51.

To obtain the smallest possible water blank values, the sample must be introduced into the apparatus only after adequate freeze drying (cf. Sect. 1.2.4.4), in a completely dust-free atmosphere with constant, low humidity. These provisions are guaranteed only if the sample transfer is undertaken in a glove box in which temperature and humidity are constant. By this means, blank values of under 1.35 μg water, corresponding to 0.15 μg hydrogen, can be attained. The blank values are more constant, the more accurately one maintains the schedule of execution of the individual steps, the amounts of purging gas, the carbon disulfide concentration, and the temperature of the furnaces. If all values vary less than $\pm 10\%$, the blank value can be reproduced within 8 hr with a standard deviation $s_{20} = \pm 4$ ng H. Automation of the procedure by programmed operation will produce a substantially greater certainty (cf. Sect. 2.2.2.3.6).

Traces of oxygen in the nitrogen purging gas, as well as nitrogen oxides resulting from the combustion of nitrogen-containing samples, form sulfur dioxide in the reaction with carbon disulfide; like the SO_2 arising from combustion of sulfur-containing compounds, it reacts with excess carbon disulfide as follows:

$$SO_2 + CS_2 \rightarrow CO_2 + 3S$$

to form elemental sulfur, without affecting the water determination. To eliminate interferences from chlorine, bromine, and iodine, the inner surface of the combustion vessel is silvered to retain these elements.

2.2.2. PROCEDURES FOR CARBON AND HYDROGEN DETERMINATION

2.2.2.1. PROCEDURE OF GOUVERNEUR, VAN LEUVEN, BELCHER, AND MACDONALD (6)

2.2.2.1.1. *Principle*

The sample (>30 μg), weighed in a little platinum boat, is burned in an oxygen stream. Water and carbon dioxide are frozen out in cold traps at -80 and $-196°$, respectively, and the oxygen remaining in the apparatus is pumped off. After thawing, each gas is allowed to expand into a known volume and its pressure is measured.

2.2.2.1.2. *Apparatus*

The apparatus comprises sections for oxygen purification (Fig. 30, 1–5), for decomposition (Fig. 30, 6–10), and for measuring (Figs. 31 and 32).

The oxygen needed for burning the sample is taken from a cylinder via a needle valve and passes through a mercury pressure relief flask (*1*), a bubble counter (*2*) filled with silicone oil, and a

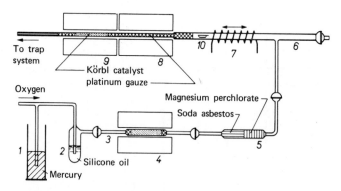

Figure 30. Decomposition apparatus for carbon and hydrogen determination after Gouverneur, van Leuven, Belcher, and Macdonald (6). *1*, pressure relief flask; *2*, bubble counter; *3*, catalyst tube for oxygen purification; *4* and *9*, electric furnaces at 500°; *5*, absorption tube; *6*, quartz combustion tube; *7*, movable heating coil; *8*, electric furnace at 1100°; *10*, platinum boat.

contact tube (*3*) filled with Körbl catalyst (*7*)[1] heated to 500°. The carbon dioxide and water resulting from the combustion of organic impurities in the oxygen are retained in the absorption tube (*5*) by a packing of soda–asbestos and magnesium perchlorate. The oxygen so purified then enters the quartz decomposition tube (*6*) in which the sample is burned in a platinum boat (*10*) by means of the movable heating coil (*7*).

The combustion products arrive with the oxygen (10 ml/min) in the hot zone of the furnace (*8*), in which the combustion is completed at 1100° over a platinum catalyst. In the final part of the decomposition tube is Körbl catalyst, heated by the furnace

---

[1] Decomposition product of silver permanganate, obtained as follows: 10 g potassium permanganate is dissolved in about 200 ml boiling water. To this solution is gradually added 10.5 g finely powdered silver nitrate. After several hours the precipitate is collected in a glass filter crucible, the solution having first been decanted and most of the precipitate retained in the precipitation vessel. The precipitate is washed 4 or 5 times with 30-ml portions of cold double-distilled water by decantation before being transferred to the filter crucible. After being dried at not above 70°, the material is heated as soon as possible, in small portions of about 2 g, in a fairly large porcelain dish until a vigorous reaction occurs. The separate fractions are collected and heated in a Supremax tube in an oxygen stream for 2 hr at about 550°. The material is stored protected from dust.

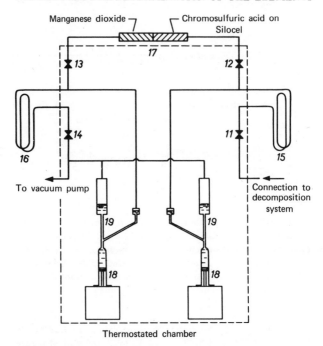

Figure 31. Measuring system for carbon and hydrogen determination after Gouverneur, van Leuven, Belcher, and Macdonald (6). *11–14*, Teflon membrane valves; *15*, cold trap for water; *16*, cold trap for carbon dioxide; *17*, absorption system for nitrogen oxides; *18*, piston burets; *19*, manometer vessels.

(*9*) to 500°, for absorption of sulfur and halogens. Steam, carbon dioxide, and excess oxygen pass through a capillary into the evacuated trap system (cf. Fig. 31) in which the water is frozen out in a cold trap (*15*) at −80° (Dry Ice–ethanol) and the carbon dioxide in a cold trap (*16*) at −196° (liquid nitrogen).

Between the two cold traps is an absorption system (*17*) for removal of nitrogen oxides, filled half with a mixture of chromic acid and sulfuric acid on SILOCEL[2] and half with manganese

---

[2] SILOCEL C_{22} of Johns Manville Corp., New York, is treated for 3 hr with a mixture of 400 ml 30% sulfuric acid a.r. and 2 ml 30% hydrogen peroxide a.r., filtered, washed free of acid, dried at 200°, and allowed to cool in a desiccator. To 250 ml of SILOCEL so cleaned is added a suspension of 3 g potassium dichromate a.r. in 50 ml concentrated sulfuric acid a.r., which has been shaken for 2 hr at 20°. This is mixed well and the SILOCEL thus prepared is stored in a tightly closed bottle in a refrigerator.

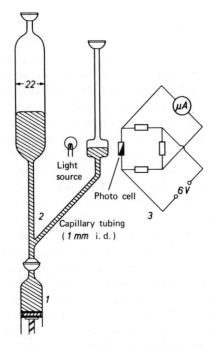

Figure 32. Mercury manometer with photometric indication of the mercury level. *1*, piston buret; *2*, manometer tubes; *3*, photometric level indicator.

dioxide. The trap system can be separated from the manometers (*19*) and the absorption system (*17*) by four Teflon valves (11–14). The entire measuring system and all valves are in a box thermostated at 50 ±0.1°. The manometers consist of U-tubes filled with mercury, connected to piston burets (Fig. 32).

When the height of the mercury column is changed by a change of pressure in the measuring leg of the manometer, it is restored to its original level, observed photometrically, by addition of mercury from the piston buret. Since the measuring leg is cylindrical, increases in pressure are proportional to the amounts of mercury added and to buret scale divisions. Changes in pressure as small as ±0.02 mm Hg can be measured in this way.

The total volumes into which the carbon dioxide and water are allowed to expand amount to 4 and 8 ml, respectively; thus 0.02 mm Hg corresponds to about 0.05 μg C per buret division and 0.02 μg H per buret division.

2.2.2.1.3   *Procedure*

At the beginning of a determination the trapping and measuring systems are evacuated by closing valve *11* (Fig. 31)) and opening valves *12–14*. Then valve *11* is opened to allow oxygen to flow into the system, and traps *15* and *16* are cooled with Dry Ice–methanol and liquid nitrogen, respectively.

Now the platinum boat containing the weighed sample is introduced and the furnace (7) (Fig. 30) turned on. After about 10 min the water and carbon dioxide will have been transferred into the cold traps; valve *11* is closed, the excess oxygen is pumped off, and after 2 min valves *12–14* are also closed. After the manometers have been zeroed, the cold trap *(15)* containing the water is heated electrically to about 75° and the carbon dioxide trap is brought to −80° by means of the Dry Ice–ethanol coolant previously used for cooling the water trap. After about 3 min the mercury levels of both measuring vessels are reset to their original positions and the volume of mercury thus used is read on each piston buret. The procedure must be calibrated in advance with various amounts of a suitable standard substance, e.g., mannitol; the carbon and hydrogen values for each sample are plotted against the respective buret readings.

While the carbon curve is linear in the region from 0 to 50 μg C, the hydrogen curve is linear only for amounts over about 2 μg H. The straight line does not pass through the origin because a certain amount of water is fixed to the vessel surface by adsorption. The major part of this water is fixed immediately upon entering the measuring system, and a minor part during the measuring process; thus an appreciable pressure drop is then observed. This interference can be taken care of, however, by keeping the observation time constant.

2.2.2.2.   PROCEDURE FOR 5–15 μg CARBON (1)

2.2.2.2.1.   *Principle*

The sample, weighed directly or by solution partition (cf. Sect. 1.2.4.4), is burned at over 1000° in pure oxygen in a quartz vessel in a quartz apparatus. The carbon dioxide formed is absorbed in 0.01M barium hydroxide solution and precipitated as barium carbonate. Barium ions remaining in solution are back-titrated against phthalein purple with EDTA in the presence of dioxane and piperidine. The endpoint is indicated photometrically.

2.2.2.2.2. *Apparatus*

Figure 33 shows the complete apparatus, comprising the gas purification section (*a*), the combustion section (*b*) (cf. Sect. 1.3.2), and the determination section (*c*). The gas purification section consists of a generator for electrolytic production of oxygen,[3] a Teflon stoppered U-tube filled with sodium hydroxide pellets, and a cold trap filled with ignited quartz wool, cooled by a Dry Ice–methanol mixture. The combustion section has already been exhaustively described in Sect. 1.3.2. It is connected to the absorption and determination section by capillary tubing (0.8 mm diameter).

The absorption vessel is equipped with a rapidly rotating bell-shaped stirrer. The stirrer, seated in a bored Teflon stopper, is introduced through the upper ⊤̄ 10 opening of the vessel and driven by a small dc motor (>4000 rpm). The bell of the stirrer is as low as possible in the lower part of the absorption vessel, so that the nozzle of the capillary from the combustion section extends into the top of the bell. The bell (12 mm height, 4 mm o.d.) has four small openings (0.7–1 mm diameter) in its lower third for gas exit, and on the upper part of the bell and lower part of the stirrer shaft there are several small nubs. At the high speed of the stirrer, the gas bubbles emerging from the bell are intensively whirled about and beaten to a froth. This results in long contact times between gas and absorbent solution. In the lower part of the vessel is a little "cuvet nose" (7) (about 1 cm^2 window surface, 3–4 mm optical path in liquid) for the photometric endpoint definition in the complexometric barium determination. On an optical bench there are a light source (low-voltage point source with constant current source), a filter (550–1000 nm transparency, e.g., OG 1 of Schott Co.), a converging lens, two collimating tubes (4 and 6 mm diameter, 100 mm length), and a photocell.[4] The cuvet nose is between the two collimating tubes. The light current is followed

---

[3] The generator (Fig. 33, *a*) consists of two separated electrode chambers with platinum foil electrodes (about 4 cm^2) and a cooling jacket. Either 2*N* sulfuric acid or 2*N* potassium hydroxide is electrolyzed. The electrolysis emf should be controllable between 0 and 48 V. The current flow is interrupted at a gas pressure of about 200 mm water.

[4] For example, photocell S60 of Dr. B. Lange Co., Berlin-Zehlendorf, Germany.

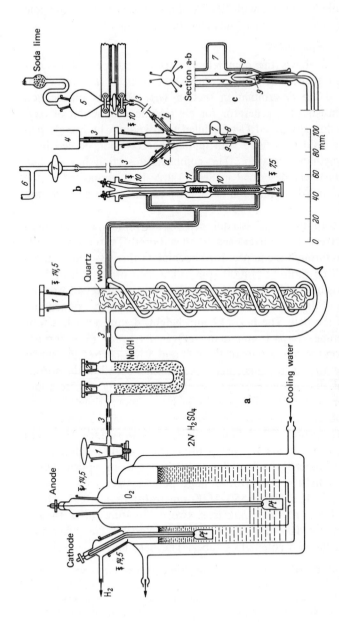

Figure 33. Complete apparatus for titrimetric ultramicro carbon determination (1). *a*, oxygen generation and purification section; *b*, decomposition section; *c*, carbon dioxide absorption and determination section; *1*, Teflon plug; *2*, Teflon stopper; *3*, polyethylene tubing connection; *4*, stirrer motor; *5*, micro piston buret; *6*, 1-ml semimicroburet; *7*, cuvet; *8*, bell-shaped stirrer; *9*, polyethylene capillary; *10*, combustion tube and holder; *11*, platinum heating coil.

with a vacuum tube voltmeter (0–10 mV) with zero suppression or with a recorder.

The indicator solution is introduced into the absorption vessel from a 1-ml microburet, the titrant solutions from two 500-μl micro piston burets with reservoirs[5] (cf. Sect. 1.2.4.2), via polyethylene tubes (1.5–1.8 mm diameter), and through three $\overline{\text{S}}$ 5 capillary joints. The solutions are carried further within the vessel by polyethylene tubes.

The capillaries conveying the titrant solutions extend to just above the vessel bottom; the indicator solution capillary should not dip into the solution to be titrated. Water can be introduced and drawn off through a fourth opening (not shown in Fig. 33), thus permitting reproducible rinsing (cf. Fig. 42).

2.2.2.2.3. *Reagents*

Approximately 0.01M barium hydroxide solution. 3.15 g Ba$(OH)_2 \cdot 8H_2O$ a.r. is dissolved in 1000 ml freshly boiled, double-distilled water in a volumetric flask previously flushed with CO_2-free air. The solution is stored in a polyethylene bottle which has also been flushed with CO_2-free air before transfer of the solution. After standing awhile, a slight precipitate of $BaCO_3$ may settle; this will have no effect on the determination, however, since the strength of the solution is to be determined daily.

0.01M EDTA solution (*disodium salt of ethylenediamine-tetra-acetic acid*) *containing 20% dioxane.* The solution must be prepared with CO_2 excluded as much as possible. Freshly boiled double-distilled water and freshly distilled dioxane are used, collected in a vessel flushed with CO_2-free air. The solution is stored in a quartz bottle previously flushed with CO_2-free air. The solution may be removed only with a transfer pipet from which the air has previously been expelled by pure nitrogen. While the pipet is dipping into the solution, a slow stream of nitrogen is fed through it to largely prevent introduction of carbon dioxide from the air.

Indicator solution. 5 ml freshly distilled piperidine, 10 ml double-distilled water (CO_2-free), and 8.00 mg phthalein purple a.r. are diluted to volume in a 100-ml volumetric flask with freshly

---

[5] The burets are closed with soda lime tubes to prevent interference from atmospheric carbon dioxide.

distilled dioxane. As in the preparation of the previous solutions, care must be taken to introduce as little atmospheric carbon dioxide as possible.

Mixture of Körbl silver–manganese catalyst (7) v' and lead chromate (sintered, for elemental analysis) in the ratio of *1:1.5*

2.2.2.2.4. *Procedure*

The cleaned apparatus should be dried well. The sample, weighed into the combustion tube[6] directly or by solution partition, is introduced into the combustion chamber of the apparatus (Fig. 33, *b*). The apparatus is then flushed with 100 ml oxygen (10 ml/min) and thereafter the oxygen flow is not interrupted during the entire determination. For the region of 5–20 μg carbon, 400$\pm$0.1 μl 0.01M barium hydroxide solution is now introduced (500 μl micro piston buret) and the platinum coil is momentarily heated while the stirrer is running (about 5000 rpm). Sulfur-free samples are burned without a catalyst for about 10 sec at 1050 – 1100°; sulfur-containing samples are burned with a catalyst at 550° within 20 sec. An emf versus temperature curve for the apparatus described here is shown in Fig. 34. For apparatus with different dimensions, the curve is determined by replacing the quartz combustion tube in the platinum–rhodium heating coil by a Pt–Pt/Rh thermocouple, of which the junction is in a quartz tube (10 mm length, 2 mm diameter). The junction should be at about the center of the heating coil. The temperature dependence on applied emf is measured with an oxygen flow of 10 ml/min.

For samples of under 20 μg, 80 ml oxygen (10 ml/min) is adequate for a complete transfer of carbon dioxide. Then 1.00$\pm$ 0.01 ml indicator solution is added from a 1-ml buret (cf. Fig. 33 (*6*)). To preclude afterflow error of the buret, the exit capillary is made so narrow that the discharge rate is less than 0.5 ml/min. The reaction solution will be deep blue. The barium ions still remaining in solution can immediately be back-titrated with the

---

[6] The combustion vessel is first ignited in oxygen. For sulfur-containing samples, about 15 mg catalyst mixture is introduced before the dissolved sample is transferred into the combustion tube and this is heated to about 600° for 20 sec in an oxygen stream for removal of traces of carbon, or else a quartz reaction tube (cf. Fig. 28*a*) filled with Körbl catalyst heated to 500° is put between the combustion chamber and the absorption vessel.

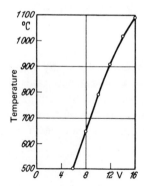

Figure 34. Relationship of temperature inside a platinum-rhodium coil to applied emf.

EDTA solution (500 μl micro piston buret). To permit solubility equilibrium it is necessary to titrate slowly. When the color of the solution has become faint, one should wait about 3 min and determine the endpoint photometrically ($\triangle$mV/0.5 μl EDTA solution) with the stirrer turned off. (The gas flow suffices for mixing.) In daylight the photometrically determined endpoint will agree with the visual endpoint. Under artificial light the visual endpoint is harder to recognize and appears somewhat too early. After the titration the contents of the absorption vessel are drawn off by the capillary provided for that purpose.

The micro piston burets are refilled and zeroed. The absorption vessel is first rinsed 5 or 6 times with a mixture of 20% acetone in double-distilled water and then twice more with pure acetone. The acetone is largely drawn off and the next determination may begin. During the cleaning and introduction of fresh absorbent solution, the oxygen flow must be so rapid that solution can never creep back into the capillary to the combustion chamber.

After about five determinations the barium carbonate deposited on the walls of the titration vessel is dissolved off with $2N$ hydrochloric acid. After about ten combustions of phosphorus-containing samples, the combustion chamber and the capillary to the absorption vessel should also be rinsed free of phosphate and well dried.

2.2.2.3. PROCEDURE FOR LESS THAN 2 μg HYDROGEN (10)

2.2.2.3.1. *Principle*

The sample, weighed directly or by solution partition (cf. Sect. 1.2.4.4), is burned in a quartz apparatus in a stream of oxygen. The water formed is converted by carbon disulfide at 500° over a corundum catalyst in a stream of nitrogen into hydrogen sulfide, which is titrated argentometrically with potentiometric endpoint indication.

2.2.2.3.2. *Apparatus*

Freeze-drying apparatus. Cf. Sect. 1.2.4.4.

Decomposition and determination apparatus. Cf. Fig. 35. The apparatus consists of a cold trap system (A), the combustion unit (B) (cf. Sect. 1.3.2), a catalyst tube with furnace (C), and the absorption and determination section (D). Parts A, B, and C are mounted in the upper half of a glove box with an air lock (cf. Fig. 3), still leaving enough room in the lower part of the box for preparations. Only the absorption section is outside the box. The nitrogen [7,8] is dried by liquid air in the two cold traps a and b and the oxygen (ordinary cylinder oxygen [8]) in cold trap c. Each Dewar flask is of 800-ml capacity. The nitrogen inlet tube branches, with one path leading to the combustion vessel B via the cold trap b, and the other path also passing through a vessel (d) containing very pure carbon disulfide over phosphorus pentoxide. The carbon disulfide temperature is kept at $0\pm0.5°$ by pumping cooled methanol through the jacket of vessel d by means of a circulating cryostat. At a constant flow rate, the nitrogen thus always assumes the same carbon disulfide concentration and the gas stream affords a reproducible blank value.

By operation of the three-way stopcock e, [9] either pure nitrogen or nitrogen with carbon disulfide vapor may be selected to pass

---

[7] Pure nitrogen is obtained from purified cylinder nitrogen which is passed through a column of BTS catalyst (BASF, Ludwigshafen) at about 150° to reduce its oxygen content (cf. Sect. 2.3.2.1.2).

[8] The gases are taken from the cylinders by two-stage reducing valves (0–1000 mm water).

[9] Teflon plug.

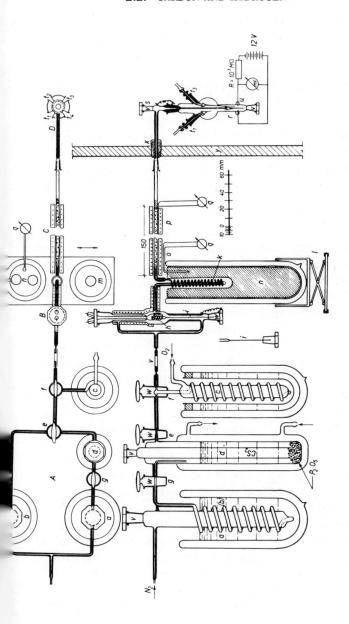

Figure 35. Complete apparatus for water determination (10). (*A*) cold trap system; (*B*) combustion unit; (*C*) catalyst chamber with furnace; (*D*) absorption and determination section; (*a, b*) cold traps with liquid air for drying nitrogen; (*c*) cold trap with liquid air for drying oxygen; (*d*) vessel with carbon disulfide over P_2O_5; (*e, f, g*) three-way stopcocks with Teflon plugs; (*h*) platinum-rhodium heating coil; (*i*) quartz combustion vessel; (*k*) quartz coil for freezing out water; (*m*) Dewar flask with liquid air; (*n*) Dewar flask with heating block; (*o*) catalyst tube, filled with α-Al_2O_3; (*p*) tube furnace; (*q*) thermocouple; (*r*) polyethylene capillary; (*s*) opening for cleaning the capillary; (*t*) $\overline{\underline{S}}$ 5 capillary joint; (*u*) silvered platinum electrode; (*v*) Teflon stopper or tube; (*w*) Teflon plug; (*x*) rubber stopper seal; (*y*) glove box wall.

through the apparatus. By appropriate setting of the three-way stopcock f^9, oxygen will arrive in the combustion chamber B via cold trap c. Stopcocks e and f can be operated through the roof of the glove box by hand or by control motors. The liquid air needed for drying the oxygen and nitrogen also provides a constant, low humidity (<0.1 g H_2O/m^3) inside the box.

The cold trap system A is connected to the combustion chamber B by a piece of teflon tubing (v). The combustion section B and the catalyst tube (o) are blown together of quartz.

The heating coil (h) in combustion section B (cf. Sect. 1.3.2) is controlled by an electromechanical time switch (cf. Sect. 2.2.2.3.6), since equal incandescence times are necessary for the attainment of constant blank values for every combustion. The quartz combustion vessel and its support (i), [10] by which the sample is brought into the platinum–rhodium coil, should have as small a surface area as possible. The combustion vessel and the inside of part B are silvered to retain halogens after the decomposition. A standard procedure, (e.g., ref. 24) is used for silvering. The treatment is repeated 4 or 5 times until there is a dense silver mirror. This is then rinsed with copious amounts of water.

The outer, upper part of B and the connecting piece to the coil (k) are tightly wound with 1-mm silver wire for heat conduction and have asbestos twine wound over this for thermal insulation. The internal diameter of the connecting capillary and the coil is 1 mm.

Under the coil (k) is an elevating platform (l), running on tracks, on which two Dewar flasks (m,n) [11] are mounted. The coil (k) can be brought into the mouth of either flask by back-and-forth and up-and-down movement of the platform. These movements can be performed manually or, better, by two geared motors.

Dewar flask m contains liquid air for freezing out water. The coil should not be cooled directly, but only through an air jacket. In Dewar flask n is an aluminum block with two holes. The coil (k) fits into one hole and the other contains a 100-W heating cartridge, enabling the block to be held at a temperature of $350\pm10°$ by

---

[10] Combustion vessel: 9 mm height, 2 mm i.d., 2.8 mm o.d.; stem: 60 mm height, 0.5–0.6 mm diameter.

[11] 300-ml capacity.

means of an autotransformer. [12] To the coil (k) is connected the catalyst tube (o), [13] surrounded by a small tube furnace (p). [14] The furnace temperature is measured with a thermocouple (q) and controlled at $500\pm10°$.

The catalyst tube (o) is filled with a mixture (1:1) of sintered corundum of 1-mm particle size (K 25 DEGUSSA, Hanau) and aluminum oxide powder (after Brockmann for column chromatography) which has been ignited for 2 hr at $1400°$. The 100-mm long charge is held fast by a 10-mm quartz wool plug at either end. The bed must allow a gas flow of 10–30 ml/min at a pressure of 200 mm water.

The catalyst tube (o) is connected to the absorption and determination vessel (D) by a $\overline{\mathcal{S}}$ 5 joint. The capillary tube passes out of the glove box through a rubber-stopper packing (x) in the wall (y). The capillary is bent down at a right angle and goes into a polyethylene capillary (r) (about 0.1 mm i.d.), which ends just above the floor of the absorption vessel. No absorbent solution should creep back into the polyethylene capillary as the gas bubbles into the solution; if this should happen the capillary must be drawn out finer. Joint s permits cleaning the inside of the capillary with dilute hydrochloric acid and water.

Zinc sulfate absorbent solution, EDTA solution, and silver nitrate titrant solution are introduced into the absorption vessel through polyethylene tubes (2 mm o.d.), capillary joints ($\overline{\mathcal{S}}$ 5), and polyethylene capillaries. The zinc sulfate and EDTA solutions are taken from 1-ml microburets with closed reservoirs and stopcocks with Teflon plugs or, in a programmed system, from the magnetic valve burets described in Sect. 1.2.4.2. The silver nitrate solution is measured out with a 500-μl piston buret with reservoir, three-way stopcock, and micrometer screw (cf. Sect. 1.2.4.2) or with a motor-driven piston buret. The glass body of the buret must be lacquered black to protect the silver solution from light.

Two platinum wire electrodes (0.5 mm length, 0.3 mm di-

---

[12] For example, Electrohahn of Voigt und Haeffner Co., Frankfurt-Berlin, Germany.

[13] 180 mm length, 3 mm i.d., 6 mm o.d.

[14] A piece of copper tubing (120 mm length, 15 mm o.d., 7 mm i.d.) is inserted into a 200-W heating element and the whole unit wrapped with asbestos tape.

ameter), silvered as described in Sect. 2.7.3.2.3, are fused into the absorption vessel 5 mm above the floor. The electrical measuring system is shown in Fig. 35 (cf. also Sect. 2.7.2.2).

2.2.2.3.3. *Reagents*

Carbon disulfide. 200 ml carbon disulfide (for spectroscopic purposes) is vigorously shaken (shaking machine) for about 2 hr with 5 ml mercury a.r., separated from the mercury, and stored in the dark over about 20 g phosphorus pentoxide a.r. in a $\overline{\$}$ bottle.

0.1 M zinc solution. 28.8 g $ZnSO_4 \cdot 7H_2O$ a.r. and 13.6 g sodium acetate ($CH_3COONa \cdot 3H_2O$) a.r. are first dissolved in some double-distilled water in a 1000-ml volumetric flask. The solution is then diluted to volume with double-distilled water and, after about 2 hr, filtered into a polyethylene bottle which is then tightly closed. The solution will keep for several months.

0.25 M EDTA, 0.5 N sodium hydroxide, and 2 N ammonia solution. 20 g sodium hydroxide a.r. is first dissolved in 100 ml double-distilled water in a 1000-ml volumetric flask. To the cooled solution is added 93 g EDTA (disodium salt) a.r. and 200 ml 10 *N* ammonia water (0.928 sp. gr. at 20°), freshly prepared by introduction of cylinder ammonia into double-distilled water. The solution is diluted to 1000 ml with freshly prepared double-distilled water and kept in a tightly closed polyethylene bottle.

0.01 and 0.005 N silver nitrate solutions. 2 ml glacial acetic acid is added to 10.0 or 5.00 ml 0.100 *N* silver nitrate solution in a 100-ml quartz volumetric flask. The solution is diluted to volume with double-distilled water and stored in the dark. It will remain titer constant for several weeks. 1 μl solution corresponds to 10.08 or 5.04 ng H.

5% potassium cyanide solution. In a polyethylene wash bottle.
2 N sulfuric acid a.r. In a polyethylene wash bottle.

2.2.2.3.4. *Apparatus Preparation* (cf. Fig. 35).

Before assembly of the apparatus, components *A* and *B–C* (without the plastic parts) are heated about 2 hr in a drying oven at 250°. When part *A* has cooled to about 100°, it is quickly connected to the oxygen and nitrogen supplies, the plastic parts are replaced, the warm cold traps are put into the Dewar flasks containing liquid air [15] in the glove box, and nitrogen and oxygen are admitted at about 20 ml/min. Part *A* is then connected to quartz section *B–C*,[16] still at about 100°, the furnace *p* is slid over the catalyst tube *o,* and part *D* is connected. After putting carbon disulfide and phosphorus pentoxide into vessel *d,* the cooling of that vessel is begun [17] and the combustion chamber (*B*), the coil (*k*), and the catalyst tube (*o*) are heated to the desired temperature while being flushed with pure nitrogen. The apparatus is heated for about 24 hr. The furnaces *p* and *n* are kept on during a measuring period. After heating, CS_2–N_2 mixture (10 ml/min) is passed through the apparatus, with the combustion chamber *B* heated to only 300–400°, until a constant blank value of less than 1.5 μg water has been attained in each determination (cf. Sect. 2.2.2.3.3). Depending on the skill in assembly of the apparatus, this may take 2–4 days more.

2.2.2.3.5. *Procedure*

The sample is weighed directly or by solution partition (cf. Sect. 1.2.4.4). In the latter case, the aliquot portion (10–20 μl) is transferred to the combustion vessel on the holder (*i*) in a glove box. The sample should contain about 0.5–3 μg hydrogen. The combustion vessel with holder (*i*) is removed through the air lock of the glove box in a microdesiccator (cf. Fig. 21) and connected to a freeze-drying apparatus (cf. Sect. 1.2.4.4). The solution is first degased carefully and the solvent evaporated in the combustion vessel under mild vacuum.[18] Drying then takes place at $<5 \times 10^{-2}$ torr for a length of time that depends on the efficiency of the pump. Generally a drying time is adequate which in prelimi-

---

[15] From this point on, the Dewar flasks *a, b* and *c* must for the duration of a measuring period be adequately supplied with liquid air through the roof of the glove box. They consume about 10 liters in 24 hr. In replenishing the liquid air in Dewars *a* and *b,* stopcock *e* must be set to block the path to *B* to prevent carbon disulfide from flowing back into traps *a* and *b.*

[16] The ⚊ 10 joint of the heating coil holder should be greased with a little silicone grease.

[17] Since the interior temperature of the glove box will be 40–50°, the cooling of vessel *d* must always function dependably. If the cooling failed, there would be danger of explosion.

[18] 10–30 torr for water, 100–150 torr for acetone or ethanol.

nary trials with pure solvent has led to constant blank values.[19]

The holder (i) containing the sample is removed from the desiccator in a glove box [20] and introduced into the combustion chamber B, with the upper half of the little combustion vessel extending into the platinum–rhodium coil (step 1 in the schedule, Table 5).

The schedule (Table 5) is the result of numerous experiments under the most varied conditions; its object is the execution of a determination in the shortest possible time. The data must be corrected for apparatus of other dimensions.

The elemental sulfur formed by reaction of carbon disulfide with traces of oxygen, nitrogen oxides, or SO_2 (cf. Sect. 2.2.1) will deposit at the exit of the catalyst tube (o) from the furnace (p) and should be removed from time to time. To do so, the absorption section D is removed and the furnace (p) is moved over the joint of the contact tube (o). The sulfur will burn and the resulting SO_2 can be led out of the glove box by a glass tube through the wall opening (x).

2.2.2.3.6. *Programmed Control*

With programmed execution of the schedule, all manual operations can be reduced to a minimum, thus largely obviating "personal errors" and attaining a constant time for each step, which is essential for good blank value constancy.

An electromechanical program system (25), with ten different control units switched in series or parallel, has been found especially suitable; each switching step takes place via a relay after a preselected number of photoelectrically generated emf pulses (1 pulse/0.6 sec) through an electromechanical counter.

According to the schedule (Table 5), the relays control the stopcocks and motive devices of the Dewars via geared motors, the heating coil, the two magnetic valve vessels, and the motor-driven micro piston buret (cf. Sect. 1.2.4), which is coupled to the paper transport of a millivolt recorder.

---

[19] Drying time about 3 hr.

[20] A tube of phosphorus pentoxide is connected to the stopcock of the desiccator before opening the desiccator.

Table 5. Determination Schedule

Step	Time, min	Gas	Gas flow, ml/min	Coil h	Coil k	Vessel D
0 Introduce sample	0	N_2	10	Off	Room temp.	Rinse with H_2O
1	0	O_2	5	Off	Dewar m (−170°)	Treat with KCN solution
2	2	O_2	5	On	Dewar m (−170°)	Rinse with H_2SO_4
3	3	N_2	10	On	Dewar m (−170°)	Rinse with H_2O
4	4	N_2	10	Off	Dewar m (−170°)	Introduce 0.25 ml Zn solution[a]
5	9	CS_2–N_2	10	Off	Dewar m (−170°)	
6	11	CS_2–N_2	10	Off	Dewar n (350°)	
7	19	N_2	10	Off	Dewar n (350°)	Add 0.25 ml EDTA solution[a]
8	27	N_2	10	Off	Dewar n (350°)	
9	29	N_2	10	Off	Room temp.	Begin titration[b]
10	32–35	N_2	10	Off	Room temp.	End titration
1 Introduce sample	0	N_2	10	Off	Dewar m	Rinse with H_2O

[a] From a 1-ml microburet with reservoir.

[b] From a 500-μl piston buret, lacquered black.

2.2.3. REFERENCES

1. K. H. Ballschmiter and G. Tölg, *Z. Anal. Chem.*, **203**, 20 (1964).
2. W. J. Kirsten, *Z. Anal. Chem.*, **181**, 1 (1961).
3. W. J. Kirsten, K. Hozumi, and L. Nirk, *Z. Anal. Chem.*, **191**, 161 (1962).
4. K. H. Ballschmiter and G. Tölg, unpublished.
5. C. W. Ayers, R. Belcher, and T. S. West, *J. Chem. Soc.*, **1959**, 2582.
6. P. Gouverneur, H. C. E. van Leuven, R. Belcher, and A. M. G. Macdonald, *Anal. Chim. Acta*, **33**, 360 (1965).
7. J. Körbl, *Mikrochim. Acta*, **1956**, 1705.
8. W. J. Kirsten, *Mikrochim. Acta*, **1964**, 487.
9. H. C. E. van Leuven and P. Gouverneur, *Anal. Chim. Acta*, **30**, 328 (1964).
10. G. Tölg and K. H. Ballschmiter, *Microchem. J.*, **9**, 257 (1965).
11. K. Abresch and J. Claassen, *Die Coulometrische Analyse*, Verlag Chemie, Weinheim/Bergstr, 1961, p. 131.
12. G. Schwarzenbach, *Die Komplexometrische Titration. Die Chemische Analyse*, Vol. 45, Verlag Ferd. Enke, Stuttgart, 1957.
13. S. T. Abrams and V. N. Smith, *Anal. Chem.*, **34**, 1129 (1962).
14. R. G. Armstrong, K. W. Gardiner, and F. W. Adams, *Anal. Chem.*, **32**, 752 (1960).
15. E. Barendrecht, *Anal. Chim. Acta*, **25**, 402 (1961).
16. E. Bönisch, *Materialprüfung*, **4**, 247 (1962).
17. H. Dirscherl and F. Erne, *Mikrochim. Acta*, **1962**, 794.
18. H. S. Knight and F. T. Weiss, *Anal. Chem.*, **34**, 749 (1962).
19. H. Malissa, *Mikrochim. Acta*, **1960**, 127.
20. D. A. Otterson, *Anal. Chem.*, **33**, 450 (1961).
21. S. Greenfield and R. A. D. Smith, *Analyst*, **87**, 875 (1962).
22. S. Mlinkó, *Mikrochim. Acta*, **1962**, 638.
23. E. Terres and H. Wesemann, *Angew. Chem.*, **45**, 795 (1932).
24. G. Brauer, *Handbuch der Präparativen Anorg. Chemie*, Verlag Ferd. Enke, Stuttgart, 1954, p. 765.
25. H. Boer, *J. Sci. Instr.*, **40**, 121 (1963).
26. R. Belcher, G. Dryhurst, A. M. G. Macdonald, J. R. Majer, and G. J. Roberts, *Anal. Chim. Acta*, **43**, 441 (1968).

2.3. OXYGEN

2.3.1. DECOMPOSITION AND DETERMINATION, GENERAL

Of the two possible methods for samples in the 10-μg region, catalytic hydrogenation of oxygen to water after ter Meulen (1) and the conversion of oxygen to carbon monoxide after Schütze (2), Zimmermann (3), and Unterzaucher (4–6), only the latter method has so far been used by Campiglio (7) for the 10-μg region.

The iodine liberated in the oxidation of carbon monoxide by iodic acid anhydride is titrated after Leipert (8) with 0.01 N sodium thiosulfate solution. Here 1 μl of 0.01 N thiosulfate solution corresponds to 0.0668 μg oxygen. For sample weights of 20–60 μg, a maximum absolute error of $\pm$0.23% is given (15 different samples). Since only single determinations were carried out, no standard deviation can be given for the procedure. The blank value is said to amount to a maximum of 0.6 μg oxygen and, in spite of the relatively large surface of the microapparatus used (9), to be very constant.

The low blank value is attained by the following means: The nitrogen serving as carrier gas is passed over BTS catalyst[21] at room temperature to remove oxygen, and over Anhydrone[22] and phosphorus pentoxide pumice for further purification. The pyrolysis tube consists of a special Brazilian kind of quartz, which after hydrofluoric acid treatment gives only a very small blank value.[23] The purity requirements of the carbon used in the conversion are fulfilled by ashless gas black CK/3 of DEGUSSA Co., Frankfurt/Main. For the quantitative conversion of the oxygen to carbon monoxide, a carbon bed of 180 mm length in a tube of 9 mm diameter and a nitrogen flow of 10 ml/min are used for 3-mg samples. The same conditions are retained for the 10-μg region; a further

---

[21] BASF, Ludwigshafen/Rhein, Germany.

[22] Magnesium perchlorate preparation of J. T. Baker Chemical Co., Phillipsburg, New Jersey.

[23] Schöniger (10) has found that quartz tubes of Heraeus Co., Hanau, Germany also produce only very small blank values.

75

decrease in blank values thus seems possible by suiting the tube dimensions to the hundred-fold smaller sample. The optimal reaction temperature of 1120° is maintained constant to ±1°. Basic and acidic pyrolysis products and hydrocyanic acid are removed before the oxidation of the carbon monoxide by Ascarite and phosphorus pentoxide on pumice. HI_3O_8, which reacts more selectively than I_2O_5 with CO (11), is used for the oxidation at 120°. It is essential that the components of the apparatus, with only one exception, be connected by ground-glass joints. The oxygen permeability of rubber tubing, especially of silicone rubber, is referred to by another author (10).

Blank values from oxygen adsorbed on the platinum boat or adsorptively bound water are kept small and constant by using boats as small as possible with equal surface areas. The boats should be briefly ignited in the reducing flame of a Bunsen burner before use. Heating in a hydrogen stream (12) is not necessary. The determination of the iodine after Leipert does not cause any difficulties.

2.3.2.   PROCEDURE FOR OXYGEN DETERMINATION

2.3.2.1.   PROCEDURE OF CAMPIGLIO

2.3.2.1.1.   *Principle*

The sample (>20 μg) is pyrolyzed at 900° in pure nitrogen in a quartz tube; the bound oxygen is converted by gas black at 1120 ±1° to carbon monoxide, which is oxidized to CO_2 by HI_3O_8 at 120°. The iodine liberated in the reaction is absorbed in 20% sodium hydroxide solution and titrated iodometrically after Leipert.

2.3.2.1.2.   *Apparatus*

Microburet after Gorbach (13) or micro piston buret. Cf. Sect. 1.2.4.2.

Determination apparatus. The apparatus consists of the nitrogen purification section (Fig. 36) and the pyrolysis and oxidation section (Fig. 37).

Nitrogen purification. Postpurified nitrogen (99.98%) is taken

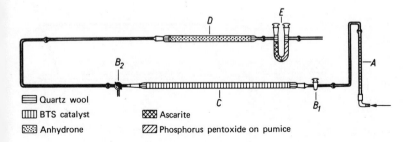

Figure 36. Nitrogen purification section of the Campiglio oxygen determination apparatus. *A*, rotameter; *B₁, B₂*, glass stopcocks; *C*, tube with BTS catalyst; *D*, tube with Anhydrone; *E*, U-tube with phosphorus pentoxide on pumice.

from the cylinder via a reducing valve and a precision needle valve[24] in a steady stream of 10 ml/min and passed through a rotameter (A)[25], through the stopcock (B_1), and through tube C which has been filled with BTS catalyst pellets of about 1 mm diameter and heated to 150°. Spent catalyst can be regenerated by hydrogen at 120°. Carbon dioxide and water are removed from the nitrogen by the absorption tube (D) filled with Anhydrone[22] and the following U-tube filled half with Ascarite and half with phosphorus pentoxide on pieces of pumice.

Stopcock B_1 serves to close off the apparatus; stopcock B_2 is used in regenerating the BTS catalyst with hydrogen, in changing the nitrogen cylinder, and in changing the pyrolysis tube. The conical joints are greased with stopcock grease and the ball joints with sealing material after Krönig (fusion of one part white beeswax and four parts rosin).

---

[24] For example, Tale regolatore è fornibile dalla Società Italiana Manometrie, Milano, Italy.

[25] For example, rotameter tube (no. 225836/16) of Rotameter Manufacturing Co., Ltd., Purley Way, Croydon, England.

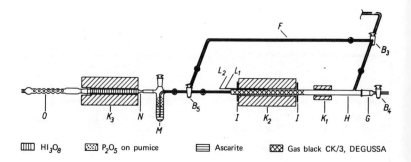

$\boxed{\text{IIIII}}$ HI$_3$O$_8$ $\boxed{\text{:::}}$ P$_2$O$_5$ on pumice $\boxed{\equiv}$ Ascarite $\boxed{\text{XXX}}$ Gas black CK/3, DEGUSSA

Figure 37. Decomposition and reaction section of the Campiglio oxygen determination apparatus. B_3, B_4, B_5, glass stopcocks; F, countercurrent flushing system; G, pyrolysis tube cap; H, pyrolysis tube; K_1, movable furnace; K_2, catalyst furnace; K_3, furnace for heating the oxidation tube; I, asbestos sheets; L_1, L_2, thermocouples; M, absorption tube, upper half filled with phosphorus pentoxide on pumice, lower half with Ascarite; N, oxidation tube with HI$_3$O$_8$; O, absorption tube.

Pyrolysis section. The purified nitrogen passes through the three-way stopcock (B_3) (Fig. 37) into the pyrolysis tube (H)[26] either directly or countercurrent, depending on the setting of stopcock B_3. The wide portion (9 mm i.d.) of the tube is about 350 mm long, the narrow portion (2 mm i.d.) 70 mm. The 180-mm gas black[27] bed should not have any channels; it is bounded at either end by a 1-cm quartz wool plug and a 3-cm layer of quartz splinters. The tube furnace K_2 should heat the carbon charge evenly to $1120\pm1°$. The temperature is controlled by an automatic controller with temperature indication (14),[28] of which both thermocouples (Pt/Pt–Rh) are introduced into the center of the furnace.

---

[26] New quartz tubes should be treated about 1 hr with 40% hydrofluoric acid, rinsed well, and dried.

[27] Gas black beads CK/3 of DEGUSSA Co., Frankfurt/M., Germany which should be ignited 4–5 hr at 1120° in a nitrogen stream in a quartz tube before use.

[28] For example, CAME Co., Via Bellosio, 15, Milano, Italy.

The small tube furnace K_1, driven by a geared motor, can be moved back and forth over the 8-cm empty part of the pyrolysis tube at a rate of 1 cm/min. Its purpose is to decompose at 900° the sample introduced into the pyrolysis tube through joint G. The current supplies of the furnaces (K_1,K_2) must be mutually independent.

The pyrolysis tube (H) is connected via stopcock B_5 to the absorption tube (M) which is filled, from bottom to top, with Ascarite, quartz wool, and phosphorus pentoxide on pumice. The absorption tube (M) is connected by a silicone rubber tube to the oxidation tube (N), which is kept at 120° by furnace K_3. The tube is carefully filled with HI_3O_8 so that there are no channels. The tube end adjoining M is narrowed to a capillary to prevent back-diffusion of iodine from N to M.

Iodine formed in the reaction of HI_3O_8 with CO is transported by the nitrogen stream into absorption vessel O, filled with 20% sodium hydroxide solution. The joint between O and N must be completely dry when the two tubes are connected and must be inside furnace K_3 to prevent iodine losses during transfer.

2.3.2.1.3. Reagents

Postpurified nitrogen
Iodic acid anhydride after Unterzaucher (11). Iodic acid anhydride a.r. is dissolved in 60% nitric acid; the crystallized HI_3O_8 is separated and dried in the dark in a desiccator over phosphorus pentoxide and potassium hydroxide.
20% sodium hydroxide solution a.r.
Bromine – acetic acid – potassium acetate solution. 100 g potassium acetate a.r. is dissolved in 1000 ml glacial acetic acid and 4 ml bromine a.r. added. The solution is stored in a brown glass-stoppered bottle.
Formic acid, 98 – 100%, a.r.
15% sulfuric acid
Potassium iodide a.r.
Thyodene (Purkins, Williams Ltd., London). 40% aqueous solution of soluble starch.
0.01N potassium iodate solution: 0.3249_5 g potassium hydrogen

iodate, $KH(IO_3)_2$ a.r., is dissolved in some double-distilled water in a 1000-ml volumetric flask; the solution is then diluted to volume with double-distilled water.

0.01N sodium thiosulfate solution. 2.482 g $Na_2S_2O_3 \cdot 5H_2O$ a.r. is dissolved in some double-distilled water in a 1000-ml volumetric flask, 2 ml amyl alcohol is added as stabilizer, and the solution is diluted to volume. This solution is kept in a brown bottle connected to a 5-ml piston buret.[29]

The thiosulfate solution is calibrated against the 0.01N potassium iodate solution as follows: With a calibrated pipet, 5.00 ml of the 0.01N potassium iodate solution is introduced into a 100-ml titration vessel. 25 ml double-distilled water, 2 ml 15% sulfuric acid, and about 150 mg potassium iodide are added, and this is titrated, while stirring with a magnetic stirrer, with the thiosulfate solution until the solution is only pale yellow. Then 2 drops 40% Thyodene solution is added and the solution titrated to incipient decoloration.

2.3.2.1.4. *Apparatus Preparation*

After the apparatus has been assembled, nitrogen is passed through the catalyst tube countercurrently for 15 min; furnace K_2 is then brought to 1120° and the catalyst tube bed heated for about 48 hr with a nitrogen flow of 10 ml/min. Meanwhile an oxidation tube (N), freshly filled with HI_3O_8, is heated in a drying oven at 150° for about 12 hr; it is then brought while still hot into furnace K_3, which is at 120°, and immediately connected to the apparatus. The apparatus is then flushed one more hour with nitrogen (10 ml/min) before determination of the apparatus blank. When the blank value has become constant, which should not require more than 9 μl thiosulfate solution, the actual determination may begin.

2.3.2.1.5. *Determination Procedure*

The apparatus is flushed countercurrently with nitrogen for 10 min (10 ml/min), the cap (G) removed (cotton gloves), the platinum boat containing the sample (20 – 60 μg) introduced into the pyrolysis tube (H) up to the quartz wool plug (steel forceps), and the decomposition tube closed with the cap, carefully

---

[29] For example, Metrohm Co., Herisau, Switzerland.

greased with Vaseline, with the stopcock (B_4) open. The cap is fastened to tube H with two steel springs. After 3 min more of nitrogen flushing, the absorption tube (O), loaded with 1 ml of 20% sodium hydroxide solution, is connected to N. The tubes M, N, and O are flushed 10 sec more with nitrogen through F, and then the stopcocks are set for the normal flushing path. Furnace K_1, placed about 1 cm before the boat, is turned on and moved at a rate of 1 cm/min. In about 4 min it will reach the catalyst furnace K_2; after 7 min it is turned off, and removed after 8 min. During the seventh minute a cooling trough is filled with Dry Ice and set on the contact tube in place of the furnace to cool it quickly. The absorption tube is removed and its contents rinsed into a 25-ml titration beaker with 10 ml distilled water. To the solution is added 500 μl bromine – acetic acid – potassium acetate solution (1-ml microburet with reservoir). The solution is carefully rotated, a drop of formic acid is added, the solution is again rotated, the fog is driven out of the beaker with air, and the solution is left standing for 3 min. While a new sample is being decomposed, the titration is carried out and the joint of the absorption tube simultaneously dried in a metal drying block. For the titration with $0.01N$ thiosulfate solution, a microburet after Gorbach is used or one of the piston burets described in Sect. 1.2.4.2. 2 ml 15% sulfuric acid and about 50 mg potassium iodide are added, the beaker rotated briefly, the buret tip dipped into the solution, and the solution titrated with stirring with $0.01N$ thiosulfate solution. Near the end of the titration a drop of 40% Thyodene is added. The entire determination takes about 13 min.

The oxygen content of the sample is calculated by

$$\% \ 0 = \frac{(V_1 - V_2) \cdot 6.667}{E}$$

where V_1 is the $0.01N$ thiosulfate solution consumption, V_2 is the blank consumption in μl, and E is the sample weight in μg.

It is interesting to note that in the analysis of fluorobenzoic acid Campiglio found no positive errors caused by fluorine.

2.3.3. REFERENCES

1. H. ter Meulen, *Rec. Trav. Chim.*, **51**, 509 (1922).
2. M. Schütze, *Z. Anal. Chem.*, **118**, 241 (1939).
3. W. Zimmermann, *Z. Anal. Chem.*, **118**, 258 (1939).
4. J. Unterzaucher, *Ber. Deut. Chem. Ges.*, **73**, 391 (1940).
5. J. Unterzaucher, *Mikrochemie (Wien)*, **36/37**, 708 (1951).
6. J. Unterzaucher, *Bull. Soc. Chim. France*, **20**, 76 (1953).
7. A. Campiglio, *Mikrochim. Acta*, **1964**, 114.
8. T. Leipert, *Mikrochem. Pregl Festschr.*, **1929**, 266.
9. A. Campiglio, *Il Farmaco, Ed. Sci.*, **19**, 385 (1964).
10. W. Schöniger, *Mikrochim. Acta*, **1965**, 679.
11. J. Unterzaucher, *Mikrochim. Acta*, **1956**, 822.
12. W. J. Kirsten, *Microchem. J.*, **4**, 501 (1960).
13. G. Gorbach, *Mikrochemisches Praktikum*, Springer, Berlin, 1956.
14. A. Campiglio, *Mikrochim. Acta*, **1961**, 796.

2.4. NITROGEN

2.4.1. DECOMPOSITION AND DETERMINATION, GENERAL

For the determination of nitrogen with samples of less than 100 μg the classic determination procedures of Dumas-Pregl, Kjeldahl (1), and ter Meulen (2–4), which proved good in microanalysis, can be adapted to the small samples.

As in the determination of extremely small amounts of carbon and hydrogen, the accuracy of the procedures for determination of very small amounts of nitrogen is primarily a problem of the attainable blank value constancy (cf. Sect. 1.3).

1. With combustion methods, which produce elemental nitrogen that is to be determined gas volumetrically, nitrogen volumes of over 5 μl, corresponding to about 5.6 μg N, can be determined with sufficient accuracy by microazotometers (5–9). Even smaller gas volumes can be measured by weighing the amount of mercury corresponding to the gas volume (10,11). Assuming a standard deviation of ± 5 μg in weighing with an ordinary microbalance, then the determination of 1 μl N_2 with a standard deviation of $\pm 4 \times 10^{-4}$ μl is conceivable. In practice, however, the procedure is limited by relatively large blank value fluctuations, which already arise in the sample decomposition, and by possibilities of error in the measuring process, in which numerous manipulations and readings are necessary. Therefore a rapid increase in relative standard deviation to about 5–10% may be expected for nitrogen amounts under 2 μg, while a mean relative error of about $\pm 1.4\%$ is indicated in the region of 3–30 μg N.

2. For the determination of very small amounts of nitrogen, the decomposition after Kjeldahl is carried out in open (12,13) or sealed decomposition tubes (14–18, 49), depending on the nature of the material under investigation. For the determination of microgram amounts of ammonia or ammonium ion, there is available a number of very sensitive titrimetric (12,13,16,19–21), spectrophotometric (22–39, 50,51), and turbidimetric (49) determination procedures which can be applied after separation of the ammonia by diffusion (12,17,40,41) or microaeration (42), or directly in the decomposition solution (15,20,49).

Although there are numerous procedures for the determination of microgram and nanogram amounts of nitrogen in biological material, only two have so far been worked out for organic elemental analysis with samples of less than 100 μg.

Belcher and co-workers (15,16,43) decompose the sample in a sealed decomposition tube at 380° in the presence of a mercury (II) salt; under these conditions even heterocyclically bound nitrogen is completely converted to NH_4^+ ions. If the decomposition is preceded by a reduction with glucose or, even more effectively, with red phosphorus and hydriodic acid, then the Kjeldahl decomposition is also applicable to compounds which contain the nitrogen bound in azo, hydrazo, nitro, and nitroso groups. The ammonia formed is titrated oxidimetrically with hypobromite solution according to

$$2NH_3 + 3BrO^- \rightarrow N_2 + 3H_2O + 3Br^-$$

The procedure has been checked with 17 different compounds which also contained sulfur and halogens. From the 55 analytical results an absolute standard deviation of $\pm 0.2\%$ N was calculated. An analysis takes about 4 hr for samples with unknown nitrogen bonding type, but several analyses can conveniently be carried out side by side.

Shah and Bhatty (49) decompose the sample for 30 min in sealed decomposition tubes at 410–420° with only pure sulfuric acid and determine the ammonium (2–14 μg N) turbidimetrically with sodium tetraphenylborate at pH 2. The absolute error of the very simple and rapid procedure is indicated as $\pm 0.3\%$ N, but only seven results are given.

3. The determination of amino, amido, nitro, and heterocyclically bound nitrogen by hydrogenating digestion is successful in the region of 0.5–3 μg, corresponding to sample weights of 4–15 μg, with a standard deviation of ± 16 ng (44). The samples are pyrolyzed in hydrogen; after the gases have been passed over finely divided iron catalyst at 400° (45), the ammonia formed is titrated in the absorption vessel of the apparatus. Especially suitable for this is an iodometric determination procedure (45) previously recommended for larger samples which, adapted to the small samples, gives a standard deviation of ± 4 ng N.

The ammonia is absorbed in 500 μl 0.001N KH(IO$_3$)$_2$ solution.

This decreases the hydrogen ion concentration of the potassium iodate solution; thus upon addition of potassium iodide less iodine separates according to

$$IO_3^- + 5I^- + 6H^+ \rightarrow 3I_2 + 3H_2O$$

than without ammonia absorption. The difference in the amount of iodine separated corresponds to the amount of ammonia. The titration is carried out biamperometrically with $0.001N$ thiosulfate solution. A direct titration with sodium hypobromite solution (44,46) with biamperometric endpoint indication is more troublesome to carry out and leads to a worse standard deviation of $s_{15} = \pm 6 \ \eta g \ NH_3$.

2.4.2. PROCEDURES FOR NITROGEN DETERMINATION

2.4.2.1. PROCEDURE OF HOZUMI AND KIRSTEN (47) FOR MORE THAN 2 μg NITROGEN

2.4.2.1.1. *Principle*

The weighed sample ($>30 \ \mu g$) is decomposed at 700° in a Pyrex or Supremax tube, filled with oxygen and sealed at both ends, in the presence of metallic copper and sodium hydroxide. The copper and sodium hydroxide serve to absorb all gases contained in the digestion tube except the nitrogen; the copper first reduces nitrogen oxides to elemental nitrogen and then binds the excess oxygen. The digestion tube, drawn out at one end to a fine point, is opened under mercury by breaking off the point; the tube fills with mercury up to the volume occupied by the nitrogen in the capillary portion. The capillary is detached below the mercury meniscus from the remaining part of the digestion tube. With the capillary horizontal, the meniscus is marked, either at atmospheric (47) or at reduced (10) pressure. The accuracy can be increased substantially by reduction of the pressure. The capillary volume corresponding to the volume of nitrogen is then filled with mercury and the mercury is then weighed. Taking into account the pressure and temperature at the time of marking, this gives the desired nitrogen volume.

2.4.2.1.2. *Equipment*

Pyrex 1720 or Supremax digestion tubes[30] *for solids.* Cf. Fig. 38. Dimensions of wide part: 120 mm length, 6 mm o.d., 0.6–1 mm wall thickness; dimensions of capillary part: 60 mm length, 0.5 mm i.d. The capillary is fused to the wider part of the digestion tube beforehand.

A three-way stopcock is inserted (rubber tubing connection) for filling the tube with oxygen; this permits sealing off the tube with oxygen flowing, without a pressure buildup.

Pyrex 1720 or Supremax digestion tubes for solutions. Cf. Fig. 39. 240 mm length, 4.5 mm i.d., 6 mm o.d. To fill the digestion tubes with oxygen, stainless steel canulas (for medical syringes)

---

[30] The digestion tubes are cleaned before use by treating them on a water bath for 1 hr with glacial acetic acid and 1 hr with distilled water. They are then rinsed several times with water and dried protected from dust.

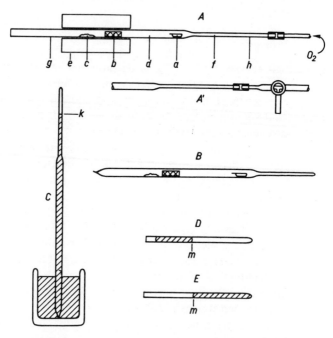

Figure 38. Course of the determination with solid samples.

are used, with 0.5 mm o.d. corresponding to the inner diameter of the drawn-out glass capillaries (cf. Fig. 39, *C*).

Tube furnace for 500°.   125 mm length.

Tube furnace for 700 – 750°.   220 mm length.

Device for marking the volume of nitrogen in the measuring capillary at reduced pressure. Cf. Fig. 40. The apparatus consists of a part with a mercury manometer (*k*), a glass stopcock (*g*), and the tube (*e*) into which the capillary is placed. The capillary is first taped onto a finely divided glass scale (e.g., of a broken thermometer), as shown in Fig. 40, *A*. A fine capillary (*h*) is connected to the stopcock (*g*) by polyethylene tubing to permit slow evacuation and readmission of air to the device. The capillary (*h*) is protected from impurities by a cotton plug (*l*).

Drying pistol. Cf. Fig. 39, *B*. To remove water and organic solvents by freeze drying, the right-hand part is set into a Dewar

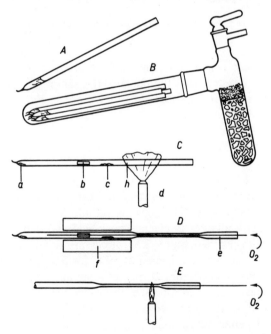

Figure 39. Course of the determination with aqueous solutions.

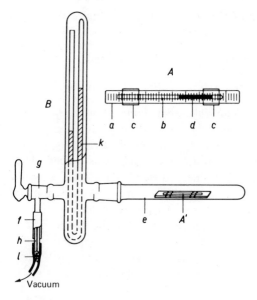

Figure 40. Device for determining the volume of nitrogen at reduced pressure.

flask containing Dry Ice–acetone and vacuum is applied. After removal of the pistol from the coolant, the sample is very quickly freeze dried in vacuum. Pieces of pumice are in the pistol to increase the surface area.

2.4.2.1.3. Reagents

Pure, nitrogen-free oxygen. Prepared by electrolysis of 2N KOH (cf. Sect. 2.2.2.2.2)

Copper gauze for elemental analysis. 0.15 mm wire diameter, 18 wires/cm². Cleaning: The gauze is cut into 50-mg pieces, which are boiled a few minutes in glacial acetic acid and then rinsed several times with distilled water. The pieces are dried by heating in a quartz tube and are then heated at 500° for 5 min in a stream of filtered air and at 800° for 20 min in a stream of pure nitrogen. After they have cooled in nitrogen, the gauze pieces are kept protected from dust in a glass-stoppered bottle. The accuracy of the procedure depends substantially on the treatment of the copper gauze.

Sodium hydroxide a.r. Solid.

Polyethylene glycol 300

2.4.2.1.4. *Determination Procedure*

2.4.2.1.4.1. *For Solids.* Between 30 and 100 μg sample (3 – 30 μg nitrogen) is weighed in an ultramicro platinum boat and placed in the horizontal digestion tube in position *a* (cf. Fig. 38, *A*). Then 50 mg copper gauze is placed at position *b* and 20 mg solid sodium hydroxide at position *c*. The digestion tube is connected to the oxygen source (3–5 ml O_2/min) via the three-way stopcock and heated to 500° in the horizontal furnace. The sample must not become hot. After 5 min the tube is removed from the furnace, the stopcock is set as shown in Fig. 38, *A'*, the tube is sealed off [31] at *g* to produce a fine point that can be easily broken, and the capillary is then closed by melting at *h*. The digestion tube, now sealed off at both ends (cf. Fig. 38, *B*), is heated over its entire length to 700 – 750° for 1–2 hours in a tube furnace. It is then allowed to cool in the furnace. All combustion products of the now completely decomposed sample, except nitrogen, will have been bound by the copper or absorbed by the sodium hydroxide.

The digestion tube is removed from the furnace, wrapped with adhesive tape from the tip of the wide part to just below the location of the sodium hydroxide, and set into a beaker of Dry Ice to freeze out all remaining water vapor on the sodium hydroxide.

The digestion tube is then immediately opened in a thick-walled vessel filled with mercury, as shown in Fig. 38, *C*, by breaking off the tip. It fills with mercury up to the part of the capillary containing the nitrogen. The capillary is quickly severed from the remaining digestion tube about 20 mm below the mercury meniscus, taped to the scale (cf. Fig. 40, *A*), and inserted into part *e* of the apparatus for volume measurement under reduced pressure (cf. Fig. 40). The nitrogen volume in the measuring capillary is increased by slow withdrawal of air with a vacuum pump until an accurate reading of the scale is assured.

With a capillary diameter of 0.5 mm, a reading accuracy of ± 0.1 mm corresponds to an absolute volume error of about ± 0.02

---

[31] This requires a hydrogen burner with a wide flame.

μl; if the volume is increased tenfold by pressure reduction, the relative error is decreased correspondingly.

The stopcock (g) is closed. After a short period for pressure stabilization, the pressure is measured on the mercury manometer (k) and the position of the mercury meniscus in the capillary is measured on the graduated support, with tube e lying exactly horizontal. Air is then admitted to the vacuum chamber through the capillary (h); the capillary and scale are removed from the tube (e) and weighed on a microbalance after removal of the mercury. Then the volume of the capillary that had been occupied by nitrogen is filled with mercury by means of a micro piston buret and the weight of the capillary is determined anew. The amount of mercury yields the volume of nitrogen at the measured vacuum and temperature. 13.5_5 mg mercury corresponds to 1 μl N_2. Converted to standard conditions, 1 μl nitrogen corresponds to 1.25_1 μg N.

2.4.2.1.4.2. *For Dissolved Samples.* The solution to be investigated is pipetted into a solution digestion tube (Fig. 39, *A*), 20 μl polyethylene glycol is added, and the solvent removed in the drying pistol (Fig. 39, *B*) by freeze drying (cf Sect. 1.2.4.4). The solution residue will be on the floor of the digestion tube. After introduction of 50 mg copper gauze and 20 mg NaOH, 80 and 100 mm, respectively, from the sample, the tube is drawn out with a wide-flame burner, 140 mm from the closed end, beyond h (cf. Fig. 39, *C*), to a capillary about 60 mm long with an internal diameter of about 0.5 mm. A steel canula is inserted through the capillary to within about 30 mm of the sample and the tube is flushed with oxygen (5 ml/min). After 5 min the canula–sample distance is increased to about 50 mm. The furnace (f) is pushed over the tube (cf. Fig. 39, *D*) with a sample–furnace distance of about 70 mm. The tube is heated to 500° for 5 min and withdrawn leftward from the furnace. The end of the steel canula is brought to the position shown in Fig. 39, *E* and the capillary immediately sealed off at the beginning of the constriction. The procedure is then as described in the previous section.

2.4.2.1.5. *Remarks*

For the 100-μg region, Hozumi (48) uses a finely powdered mixture of seven parts by weight barium oxide and one part magnesium oxide in place of sodium hydroxide and can thereby omit freezing out the remaining water vapor before filling the digestion tube with mercury. No results with this variant have been published for the 10-μg region.

2.4.2.2. PROCEDURE OF BELCHER, CAMPBELL, AND GOUVERNEUR (15) FOR MORE THAN 2 μg NITROGEN

2.4.2.2.1. *Principle*

The decomposition of the sample (40–80 μg) is conducted in a closed tube with sulfuric acid in the presence of a mercury(II) salt. Azo, hydrazo, and nitro compounds must be reduced with glucose or, preferably, with hydriodic acid and red phosphorus before decomposition. The ammonia is immediately oxidized in the digest by sodium hypochlorite solution in the presence of potassium bromide at pH 7.5–9.6. Excess oxidant is back-titrated iodometrically.

2.4.2.2.2. *Equipment*

Digestion vessels. Hemispherically closed borosilicate glass tubes (13 mm o.d., 11 mm i.d., 70 mm length). The tubes are cleaned by treating them either overnight with chromosulfuric acid or for a few minutes with a 40% hydrofluoric acid–Teepol [32] mixture (1:20), rinsed thoroughly with tap water and then with distilled water, and dried at 120°. The digestion tubes are kept in a desiccator over phosphorus pentoxide. It is essential that they dry completely and be free of contaminants (grease, *inter alia*).

Heating block. Electrically heated aluminum block (500–800 W) with several holes into which the digestion tubes fit. The block temperature should be controllable.

Centrifuge. Electric semimicro centrifuge (about 4000 rpm).

Titration equipment. Consisting of a magnetic stirrer, a 500-μl piston buret, and a daylight lamp. Also four additional 500-μl micro piston burets (cf. Sect. 1.2.4.2).

2.4.2.2.3. *Reagents*

Glucose a.r.
Hydriodic acid (density = 1.70) a.r.

---

[32] Teepol is an English Shell detergent of higher secondary alkyl sulfonates.

Red phosphorus a.r.

Concentrated sulfuric acid a.r.

2% mercury(II) sulfate solution. A few drops concentrated sulfuric acid a.r. and about 20 ml distilled water are added to 2 g mercury(II) sulfate a.r. The mixture is stirred vigorously and additional distilled water added until the residue is dissolved. The solution is then diluted to 100 ml.

2N sodium hydroxide solution

2% sodium bicarbonate solution

0.02N sodium hypochlorite solution. The solution is obtained by dilution of the necessary volume of commercial sodium hypochlorite solution (about 13–14% active chlorine). The solution must not contain any bromine and must be stored protected from light. The micro piston buret used for measuring out the solution exactly should be wrapped with black paper or lacquered black. The solution is standardized against ammonium sulfate as follows: To a weighed amount of ammonium sulfate (about 50 μg) in a titration tube are added 10 μl concentrated sulfuric acid (micro piston buret) and a small crystal of mercury(II) sulfate (about 10 μg) (platinum needle). The digestion tube is sealed off and heated in the heating block at 380° for 45 min. The further treatment is described in Sect. 2.4.2.2.4.4. The blank value is determined at the same time under identical conditions. 1 μl 0.02N sodium hypochlorite solution corresponds to 0.0934 μg nitrogen.

30% potassium bromide solution. Potassium bromide a.r. is dissolved in distilled water. The solution is stored in a brown glass bottle.

30% potassium iodide solution. The solution is prepared fresh daily from potassium iodide a.r. and distilled water and kept in a brown glass-stoppered bottle.

Approximately 4N sulfuric acid

0.01N sodium thiosulfate solution. Prepared by dilution of 0.1 N standard solution. To determine the concentration of the sodium thiosulfate with respect to sodium hypochlorite, a known amount of hypochlorite solution (e.g., 50 μl) is introduced into 1 ml water in a titration flask, then 1 drop 30% potassium iodide solution and 4 drops 4N sulfuric acid are added, and the liberated iodine is titrated as described in Sect. 2.4.2.2.4.4.

Ammonium sulfate a.r. Finely powdered and dried in a vacuum desiccator over phosphorus pentoxide. 21.19% nitrogen.

2.4.2.2.4. *Determination Procedure*

2.4.2.2.4.1. *Reduction with Hydriodic Acid and Red Phosphorus for Unknown Nitrogen Compounds.* 40–80 μg sample is weighed into a digestion tube by difference weighing. 100 μg red phosphorus and 50 μl hydriodic acid are added, the tube is heated in the heating block from 150 to 190° over a period of 30 min, and the temperature is kept at 190° for an additional 30 min. Then the tube is allowed to cool and 50 μl water and 20 μl concentrated sulfuric acid are added from 500-μl piston burets. The tube is heated in the heating block, first to 150° and then to 200° for 30 min. The greater part of the hydriodic acid is thereby evaporated. To remove the iodine and any bromine, it is then heated to 300° for 60 min. To the cooled residue is added mercury (II) sulfate solution (100 μl); the tube is then sealed off [33] and heated at 380° for 45 min in the heating block.

2.4.2.2.4.2. *Pretreatment of Samples with Amino Groups or Heterocyclically Bound Nitrogen.*   *1.* In the absence of bromine and iodine. The weighed sample is transferred into a decomposition vessel, 10 μl concentrated sulfuric acid and 10 μg mercuric sulfate are added and the digestion tube is sealed off and heated 45 min at 380°. The remaining procedure is described in Sect. 2.4.2.2.4.4.

2. In the presence of bromine and iodine. The weighed sample is transferred into a digestion tube and 20 μl concentrated sulfuric acid added. With a torch the upper part of the tube is drawn out to a constriction of only 2–3 mm. The tube is heated upright on a hotplate (300°) for 1 hr, its top covered with a watch glass. The constriction of the digestion tube should be at 100–110°. After addition of a crystal of mercury(II) sulfate ($\sim$10 μg), the tube is sealed off and handled according to Sect. 2.4.2.2.4.4.

2.4.2.2.4.3. *Pretreatment of Samples with Azo, Hydrazo, and Nitro Groups.*   With few exceptions (e.g., 2, 4-dinitrophenylhy-

---

[33] The digestion tube is held at its lower end by an asbestos-wrapped clamp; the glass about 2 cm from the top is softened by a hand-held oxygen-gas torch, the top is pulled off with forceps, and the tube is sealed off.

drazones) a reduction with glucose before the digestion is sufficient for these nitrogen compounds.

The weighed sample is transferred to a digestion tube and treated with 0.7–1 mg glucose, causing the entire sample to be on the bottom of the tube. After addition of 20 μl concentrated sulfuric acid the upper part is constricted to 2–3 mm and the open tube is heated to 300° without addition of mercury(II) sulfate.[34] After sealing, it is heated 45 min at 380°.

2.4.2.2.4.4. *General Determination.* After it cools, the digestion tube is opened by scratching it with a glass cutter about 1.5 cm below the sealed tip, wetting the scratch, and touching it with a glowing glass rod.

Both parts of the digestion tube are heated at 100° for 10 min in the heating block to drive off sulfur dioxide. Then both parts, protected from dust, are allowed to cool. After the contents of the upper part have been transferred to the main part by a capillary pipet with two 250-μl portions of water, about 100 μl mercury (II) sulfate solution is added and the walls of the tube are rinsed with about 500 μl water (capillary pipet or piston buret).

The tube is now clamped into the holder of the titration apparatus as the titration vessel, it is provided with a stirring bar, and the stirrer is turned on. Now $2N$ sodium hydroxide is carefully added from a 500-μl piston buret until the precipitating mercuric oxide just fails to dissolve.[35] A daylight lamp and a black background are recommended for better observation. For the final neutralization, an excess of about 6 drops of 2% sodium bicarbonate solution is added. After addition of 2 drops of 30% potassium bromide solution the precipitate dissolves with stirring in about 30 sec. Occasionally a slight turbidity remains which, however, does not affect the determination.

With the stirrer turned on, 200 μl 0.02 N sodium hypochlorite solution is now introduced from a 500-μl piston buret, the mixture is left standing in the dark for about 5 min for completion of the

---

[34] Mercury(II) sulfate should be omitted in this case, for its effect would be detrimental to the decomposition.

[35] The precipating mercuric oxide serves as indicator for adjusting the pH to 7.5–9.6, the region necessary for the titration; organic indicators would be oxidized by the hypochlorite.

ammonia oxidation, and the tube is then replaced in the titration apparatus. After addition of 1 drop 30% potassium iodide solution and about 250 μl (4 drops) 4 N sulfuric acid, the liberated iodine is slowly back-titrated with 0.01 N sodium thiosulfate standard solution with stirring. Toward the end of the titration some solid Thyodene [36] or soluble starch is added and the titration completed. The endpoint is observed with a daylight lamp. To determine the blank value, a parallel determination is carried out without addition of sample.

2.4.2.3. PROCEDURE FOR LESS THAN 2 μg NITROGEN (44)

2.4.2.3.1. *Principle*

The sample (4–15 μg) is measured out by solution partition. Larger samples can be weighed out directly. The sample is pyrolized at over 1000° in a hydrogen stream in a special apparatus. The nitrogen is subsequently converted to ammonia over an iron catalyst at 400°, and this is titrated iodometrically with biamperometric endpoint indication.

2.4.2.3.2. *Apparatus*

Portions of the apparatus have been described in Sect. 1.3.2. Hydrogen generated by electrolysis of 2 N potassium hydroxide solution (cf. Fig. 41, A_1) is first passed through a bubble counter (A_2) and over solid sodium hydroxide (A_3) and then, to remove traces of oxygen, passed through a small catalyst tube (A_4) (150 mm length, 4 mm i.d.) which is filled with palladium–asbestos and heated to 300–400° by two 200-W heating elements (7). The temperature is controlled by an autotransformer. Subsequently the gas stream is dried in a cold trap (A_5) (Dry Ice–methanol). The cold trap is connected to the pyrolysis chamber (Fig. 42, B_1) via two paths [three-way stopcock with Teflon plug (A_6)]. By operation of the three-way stopcock the gas stream may be passed directly, throttled by a glass stopcock with Teflon plug (A_7), or completely stopped.

Between the pyrolysis chamber (Fig. 42, B_1) and the absorption vessel (B_3) is a quartz tube (B_2) (150 mm length, 3 mm i.d.)

---

[36] Source: Purkis Williams Ltd., London.

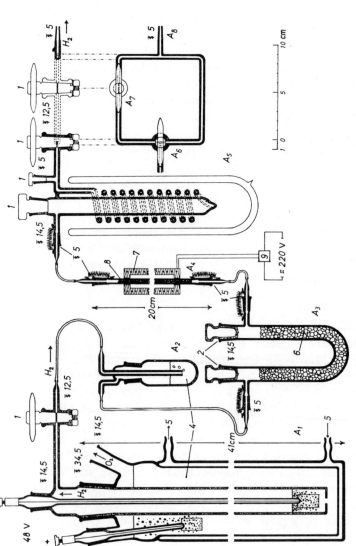

Figure 41. Hydrogen generation and purification for nitrogen determination with 4–15 µg sample (44). A_1, hydrogen generator; A_2, bubble counter with 2 N NaOH; A_3, U tube with NaOH pellets; A_4, catalyst tube filled with palladium–asbestos, heated to 300–400°; A_5, cold trap in a Dewar filled with Dry Ice–methanol; A_6, three-way stopcock for dual path; A_7, stopcock; A_8, joint for connection to the pyrolysis section (Fig. 42, B_1); 1, Teflon parts; 2, polyethylene stopper; 3, platinum foil electrodes (30 x 10 x 0.1 mm); 4, 2 N potassium hydroxide solution; 5, cooling water; 6, sodium hydroxide pellets; 7, electric furnace (two 200-W heating elements); 8, palladium–asbestos; 9, autotransformer.

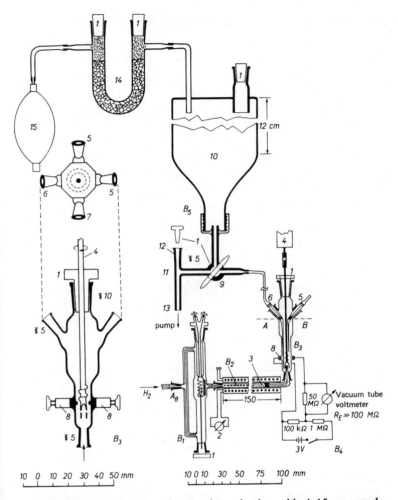

10 0 10 20 30 40 50 mm   10 0 10 30 50 75 100 mm

Figure 42. Apparatus for nitrogen determination with 4–15 μg sample. A_8, connection to apparatus for hydrogen generation and purification (cf. Fig. 41); B_1, pyrolysis section; B_2, catalyst tube filled with silver and iron powders; B_3, absorption and determination vessel; B_4, circuit for iodometric titration with biamperometric endpoint determination; B_5, rinsing system; *1*, Teflon stopper or plug; *2*, thermocouple; *3*, tube furnace (soldering iron heating element); *4*, synchronous motor and bell stirrer; *5*, polyethylene tubes to 500-μl piston burets; *6*, polyethylene tube to rinsing system (B_5); *7*, ground joint; *8*, fused-in platinum electrodes; *9*, three-way stopcock; *10*, 2-l polyethylene vessel for rinse water; *11*, tee; *12*, ground joint; *13*, connection to aspirator; *14*, U tube containing NaOH pellets and quartz wool; *15*, rubber bulb with valve.

filled half with silver powder and half with iron catalyst. Around the tube are two 150-W heating elements (*3*) wrapped with asbestos tape. The furnace must be especially well insulated at the passage to the absorption vessel, for the absorption vessel (B_3) should not be heated over 25°. The temperature of the catalyst furnace is controlled by an autotransformer and a thermocouple (Pt/Pt–Rh). The absorption and determination vessel (B_3) is similar to the type described in Sect. 2.2.2.2.2. In place of the cuvet, two platinum electrodes (*8*) (3 mm length, 0.3 mm diameter) are fused in just above the bell stirrer (Fig. 42, B_3). The thermostatable vessel described in Sect. 2.7.3.2.2, which is provided with replaceable electrodes, has proved even better. The electrical circuit for the biamperometric endpoint determination is shown in Fig. 42, B_4. The electrodes are polarized by 150 mV (voltage divider circuit). The change in polarization current over the voltage drop is followed across a 50 Mohm resistance by a high resistance vacuum tube voltmeter (0–100 mV, $R_E > 100$ Mohm) or a millivolt recorder.

The absorption vessel is connected to two 500-μl piston burets via thin polyethylene tubes (1.6 mm diameter) (*5*), of which only one is shown in Fig. 42, B_3. For a semiautomatic titration, motor-driven burets are used, one of which is coupled to the paper transport of a recorder. The micro piston burets and tubes are lacquered black for protection from light. For cleaning the vessel, a third polyethylene tube (*6*) is connected with the rinsing device (B_5) via a three-way stopcock (*9*); to this are connected a 2-liter polyethylene reservoir of double-distilled water (*10*) and, via a tee (*11*), a tube to an aspirator (*13*). By appropriate setting of the stopcock (*9*), the absorption vessel can be rinsed with water of constant pH from the bottle (*10*) by squeezing the rubber bulb (*15*). The air forced into the reservoir is cleaned by passing it through NaOH pellets and quartz wool in the U-tube (*14*). When the opening (*12*) is closed and the stopcock (*9*) reset, the rinse water can be drawn out again through the polyethylene capillary, which extends to the floor of the absorption vessel. During this withdrawal, laboratory air enters the absorption vessel through joint *7;* this air should be cleaned by passing through NaOH pellets and quartz wool. The polyethylene capillary has a hardly perceptible capillarity for water and therefore no absorbent solution will rise in it when no vacuum is applied. Thus the absorption vessel can be cleaned with water of constant NH_3 blank value between determinations without having to be opened.

2.4.2.3.3. *Reagents*

0.1 M KH(IO$_3$)$_2$ stock solution. 17.59 g iodic acid a.r., dried over P_2O_5, is dissolved in 50 ml 1.00 N potassium hydroxide solution in a 1000-ml volumetric flask and diluted to volume with CO_2-free double-distilled water. The solution is stored in the dark. A 0.001 M working solution prepared from this will remain titer constant for several weeks in a quartz or Duranglas vessel protected from light. The vessel should first be steamed. In the steamed, black lacquered reservoir of the micro piston buret the titer will change only slightly in the first few days, but will then remain constant for at least 14 days.

0.001 N sodium thiosulfate and 0.1 N potassium iodide solution. 16.6 g KI a.r. ($<0.00025\%$ $IO_3{}^-$) is dissolved in some double-distilled, CO_2-free water in a 1000-ml quartz volumetric flask. 10.00 ml 0.1 N sodium thiosulfate solution is added and the solution diluted to volume. In a black lacquered quartz bottle the titer of the solution will change only insignificantly over the first few days.

0.100 N KIO$_3$ solution

0.001 N KIO$_3$ standard solution. Prepared from the above for each use.

Approximately 0.1 N sulfuric acid a.r.

Precipitated silver powder. This is prepared by reduction of $AgNO_3$ solution by iron (II) sulfate. The preparation is rinsed well with double-distilled water and dried at 200°.

Granulated Fe$_2$O$_3$. This is prepared by ignition of Fe $(NO_3)_3$ · $6H_2O$. Sieve fraction: 20 mesh/in. (64 mesh/cm^2) to 30 mesh/in. (144 mesh/cm^2).

Lead acetate cotton. Surgical cotton is soaked in a solution of 2% lead acetate a.r. in water and dried in a drying oven at 110°.

Palladium asbestos. 10% Pd.

Ignited quartz wool

Sodium hydroxide pellets a.r.

2.4.2.3.4. *Putting the Apparatus into Operation*

The catalyst tube (Fig. 42, B_2), completely filled with Fe_2O_3 and with a quartz wool plug in its right end, is flushed (20 ml/min) at 500° with electrolytically generated hydrogen (cf. Sect. 2.4.2.3.2) until no more water condensate appears at the exit nozzle. While still being flushed with hydrogen, the tube is allowed to cool and the greatly contracted contents are gathered into the right half. If a bed of 60 mm length has not been attained, Fe_2O_3 must be added and the procedure repeated. The other half of the tube is then loosely filled with silver powder so that a hydrogen stream of at least 20 ml/min can be maintained. Replacement tubes are prepared in the same way and should be stored tightly closed. The apparatus is now assembled in final form, the iron catalyst reduced one more hour at 500°, the temperature of the catalyst furnace set to 400 ± 10°, and that of the Pt–Rh coil in the pyrolysis chamber set to 600–700°. The hydrogen flushing may now be interrupted only briefly. After about 1 hr of heating, the apparatus is ready for use. Heating of the Pt–Rh coil is stopped and the pyrolysis section is allowed to cool.

Before the first titration and upon noise in the endpoint indication, the platinum electrodes in the absorption vessel [cf. Fig. 42, B_3 (*8*)] should first be degreased with acetone and then alternately connected as cathode in dilute sulfuric acid until vigorous gas evolution can be seen on the entire electrode surfaces.

2.4.2.3.5. *Determination Procedure*

The sample, which should contain 0.5–3 μg N, is brought in the quartz combustion vessel into the Pt–Rh coil,[37] with the vessel reaching only about two-thirds into the coil. The air is displaced from the apparatus by a 5-min purge (20 ml/min) with hydrogen. Meanwhile, the two piston burets are filled with 0.001 N $KH(IO_3)_2$ solution and 0.001 N thiosulfate–potassium iodide solution from their respective reservoirs; the absorption vessel (B_3) is cleaned several times with water from the rinsing system (B_5); the

---

[37] For samples containing fluorine, directly weighed samples are loosely covered with MgO a.r. in the combustion vessel; after solution partition the combustion vessel is filled with MgO and the dissolved sample then transferred.

last rinse water is largely drawn off. After exactly 500.0 μl 0.001 N KH(IO$_3$)$_2$ solution has been introduced, the hydrogen flow is adjusted by means of the two stopcocks (A_6,A_7) on the dual path to about 5 ml/min (about 0.7 A electrolysis current at the hydrogen generator) and the Pt–Rh coil is quickly heated to over 1000°. The power consumption of the coil should be about 70 W. After about 30 sec this is reduced to 25 W. The ammonia formed is quantitatively transferred after 15 min into the absorption vessel without stirring. For the subsequent titration the hydrogen flow is stopped,[38] the stirrer is turned on at constant speed (about 2600 rpm), and the thiosulfate–KI solution is added at a rate of 2–3 μl/sec until the needle of the vacuum tube voltmeter begins to deflect. Then smaller amounts are added; 30 sec after each addition the potential change is measured. Plotted against the amount of solution added, this gives a curve similar to that in Fig. 62, which is evaluated as described in Sect. 2.9.3.2.4.2. In automatic titration, thiosulfate solution is added at a rate of 1 μl/sec.

The procedure for determination of the blank for the volume of KH(IO$_3$)$_2$ solution used, which should be determined daily, is analogous except that the combustion tube is not replaced between determinations. The amount of nitrogen in the sample is calculated from the difference between the two values, with 1 μl of 0.001 N thiosulfate solution corresponding to 0.0140$_1$ μg nitrogen.

The standard deviation of this titration amounts to $s_{20} = \pm$ 4 ng nitrogen. The thiosulfate–iodide solution is standardized against acidified 0.001 N KIO$_3$ or by hydrogenation of a standard, e.g., acetanilide or phenyl urea, and the calibration factor is checked daily. In the second standardization procedure, which takes somewhat longer, systematic errors which may arise in the procedure are also cancelled out. The two calibration procedures lead to equal results.

---

[38] At constant temperature of the platinum coil in the pyrolysis chamber and of the catalyst furnace, it is not possible for sizable amounts of solution to creep back into the catalyst tube. At worst, the gas exit capillary may fill with some solution, which toward the end of the determination is forced back into the absorption vessel.

2.4.3. REFERENCES

1. R. B. Bradstreet, *The Kjeldahl Method for Organic Nitrogen,* Academic Press, New York, 1965.
2. H. ter Meulen, *Rec. Trav. Chim.,* **43**, 1248 (1924); **49**, 396 (1930).
3. H. ter Meulen, *Bull. Soc. Chim. Belg.,* **49**, 103 (1940).
4. J. Holowchak, G. E. C. Wear, and E. L. Baldschwieler, *Anal. Chem.,* **24**, 1754 (1955).
5. J. A. Kuck and P. L. Altieri, *Mikrochim. Acta,* **1954**, 17.
6. W. J. Kirsten and B. W. Grunbaum, *Anal. Chem.,* **27**, 1806 (1955).
7. W. J. Kirsten, *Z. Anal. Chem.,* **181**, 1 (1961).
8. G. Gutbier and M. Boetius, *Mikrochim. Acta,* **1960**, 636.
9. W. Merz, *Z. Anal. Chem.,* **207**, 424 (1965).
10. W. J. Kirsten and K. Hozumi, *Anal. Chem.,* **34**, 434 (1962).
11. W. J. Kirsten and K. Hozumi, *Mikrochim. Acta,* **1962**, 777.
12. D. Bruël, A. Holte, K. Linderstrøm, and K. Rozits, *Comp. Rend. Trav. Lab. Carlsberg, Ser. Chim.,* **25**, 289 (1946).
13. E. R. Tompkins and P. L. Kirk, *J. Biol. Chem.,* **142**, 477 (1942).
14. P. Baudet and E. Cherbuliez, *Helv. Chim. Acta,* **40**, 1612 (1957).
15. R. Belcher, A. D. Campbell, and G. Gouverneur, *J. Chem. Soc.,* **1963**, 531.
16. R. Belcher, T. S. West, and M. Williams, *J. Chem. Soc.* **1957**, 4323.
17. B. W. Grunbaum, P. L. Kirk, L. G. Green, and C. W. Koch, *Anal. Chem.,* **27**, 384 (1955).
18. C. M. White and M. C. Long, *Anal. Chem.,* **23**, 363 (1951).
19. B. W. Grunbaum, F. L. Schaffer, and P. L. Kirk, *Anal Chem.,* **24**, 1487 (1952).
20. H. W. Harvey, *Analyst,* **76**, 657 (1951).
21. G. D. Christian, E. C. Knoblock, and W. C. Purdy, *Anal. Chem.,* **35**, 2217 (1963).
22. H. C. Burck, *Mikrochim. Acta,* **1960**, 200.
23. H. Roth, *Mikrochim. Acta,* **1960**, 663.
24. J. C. Mathies, P. K. Lund, and W. Eide, *Anal. Biochem.,* **3**, 408 (1962).
25. K. R. Middleton, *J. Appl. Chem,* **20**, 281 (1960).
26. W. Massmann, *Z. Anal. Chem.,* **193**, 332 (1962).
27. O. Minari and D. B. Zilversmit, *Anal. Biochem.,* **6**, 320 (1963).
28. G. Fels and R. Veatch, *Anal. Chem.,* **31**, 451 (1959).
29. A. A. Kanchukh, *Biokhimiya,* **26**, 393 (1961).

30. St. Jakobs, *Microchem. J.,* **9**, 387 (1965).
31. W. T. Bolleter, C. J. Bushman, and P. W. Tidwell, *Anal. Chem.,* **33**, 592 (1961).
32. M. K. Muftic, *Nature,* **201**, 622 (1964).
33. L. J. F. Böttcher, C. M. van Gent, and C. Pries, *Rec. Trav. Chim,* **80**, 1157 (1961).
34. L. T. Mann, *Anal. Chem.,* **35**, 2179 (1963).
35. J. T. Wearne, *Anal. Chem.,* **35**, 327 (1963).
36. P. G. Scheurer and F. Smith, *Anal. Chem.,* **27**, 1616 (1955).
37. J. Kruse and M. G. Mellon, *Sewage Ind. Wastes,* **24**, 1098 (1952).
38. T. Unemoto, Y. Truda, and M. Hyashi, *J. Pharm. Soc. Japan,* **80**, 1089 (1960).
39. F. Zitomer and J. L. Lambert, *Anal. Chem.,* **34**, 1738 (1962).
40. E. J. Conway, *Microdiffusion Analysis and Volumetric Error,* Crosby Lockwood, London, 1947.
41. C. A. Parker, *Austral. J. Exptl. Biol. Med. Sci.,* **39**, 515 (1961).
42. A. E. Sobel, A. M. Mayer, and S. P. Gottfried, *J. Biol. Chem.,* **156**, 355 (1944).
43. R. Belcher, R. L. Bhasin, and T. S. West, *J. Chem. Soc.,* **1959**, 2585.
44. G. Tölg, *Z. Anal. Chem.,* **205**, 40 (1964).
45. N. E. Gelman and M. O. Korhun, *Zh. Anal. Khim.,* **12**, 123 (1957).
46. I. M. Kolthoff, W. Stricks, and L. Morren, *Analyst,* **78**, 405 (1953).
47. K. Hozumi and W. J. Kirsten, *Anal. Chem.,* **34**, 434 (1962).
48. K. Hozumi, *Anal. Chem.,* **35**, 666 (1963); **38**, 641 (1966).
49. R. A. Shah and N. Bhatty, *Mikrochim. Acta,* **1967**, 81.
50. M. Namiki, Y. Kakita, and H. Gotô, *Talanta,* **11**, 813 (1964).
51. V. A. Konnov, *Tr. Inst. Okeanol., Akad. Nauk SSSR,* **79**, 11 1965).

2.5. SULFUR

2.5.1. DECOMPOSITION, GENERAL

Decompositions with sodium peroxide or alkali metals (1), which can be used in the 100-μg region, produce too large blank value variations in the 10-μg region. The Carius decomposition method gives satisfactory results (2, 3), but is more complicated than combustion of the sample in an oxygen flask.

To reduce the blank value, Belcher, Gouverneur, Campbell, and Macdonald (4) wrap the sample in polyethylene film. Fildes and Kirsten (5) work without combustible sample supports according to the hot flask method (6) described by Kirsten, while Tölg (7) uses specially cleaned filter paper supports with an area of 200 mm^2 as small and constant as possible (cf. p. 43 ff).

In so doing it was established (8) that uncontrollable blank values could easily arise in the treatment of the absorbent solution after the decomposition in the oxygen flask; thus it is necessary to work very carefully to attain blank value variations of less than ±20 ng. Therefore hydrogenating decomposition (9–13) is preferable (8) for the determination of less than 1 μg sulfur; the decomposition and the determination of the hydrogen sulfide formed can be carried out with laboratory air totally excluded. Wet decomposition with perchloric acid and nitric acid (14) in an open vessel proceeds quantitatively only for compounds with sulfur in the highest oxidation state (e.g., for sulfonic acids) (7); to prevent sulfur losses by evaporation the decomposition must take place in the presence of alkaline earth ions.

2.5.2. DETERMINATION, GENERAL

After decomposition in the oxygen flask the sulfur is present chiefly as sulfite, which can be oxidized to sulfate by hydrogen peroxide or bromine water in alkaline solution. Decomposition with perchloric acid–nitric acid produces sulfate directly; hydrogen sulfide is formed in hydrogenating decomposition. The reduction of microgram amounts of sulfate to hydrogen sulfide by a gas evolving procedure (7, 14, 15) is also possible. Sulfate and sulfide are therefore the recommended determination forms.

2.5.2.1. DETERMINATION AS SULFATE

The barium–Thorin procedure (16, 17) is the most accurate of all sensitive sulfate determination procedures (4, 7, 18–20). For 6 μg S per milliliter its relative standard deviation amounts to $s_{15} = \pm\ 0.7\%$. For 2 μg S per milliliter, however, the relative standard deviation is $s_{15} = \pm\ 1.5\%$. Iodide, phosphate, and several metal ions interfere with the procedure, but because of its simplicity it has been recommended for more than 3 μg sulfur (4) (cf. Sect. 2.5.3.1).

2.5.2.2. DETERMINATION AS SULFIDE

The procedures for sulfide determination (7) have greater sensitivities than the sulfate determination. The methylene blue method (15, 21) yielded a relative standard deviation of $s_{20} = \pm3\%$ for 1 μg sulfur in 2.5 ml solution. It is recommended by Fildes and Kirsten (5) for >10 μg sulfur.

Beilstein (21) obtained an inadequate standard deviation of $\pm\ 5\%$ for a titrimetric variant of the iodine–azide reaction (22, 23) with biamperometric endpoint indication for 0.9 μg S/2.5 ml.

The fluorescence methods recommended for the determination of very small amounts of sulfur, which depend on the fluorescence quenching of, e.g., fluorescein-mercuric acetate (24), are very sensitive but not very accurate (32).

The following titration procedures, on the other hand, afford sufficient accuracy when they are adapted to very small amounts of sulfur (8):

1. Iodometric titration with biamperometric endpoint determination (25).

2. Argentometric titration in a solution of sodium hydroxide and ammonia with potentiometric endpoint indication (26, 27).

3. Titration with cadmium solution, indicated by dithizone (7).

4. Direct titration with sodium hypobromite solution in alkaline solution with biamperometric indication.

These procedures presuppose working under totally oxygen-free conditions protected from light (32).

1. The sulfide oxidation must be carried out in sulfuric acid solution with an excess of 0.001 N iodine solution. The excess iodine is back-titrated with 0.001 N thiosulfate solution. The direct titration of sulfide with iodine solution will succeed in only a narrow pH range; it is very uncertain.

The standard deviation of the procedure is $s_{15} = \pm 11$ ng S for 1 μg S^{2-}/ml.

Microgram amounts of chloride, bromide, iodide, fluoride, and ammonia do not interfere. The titers of the $0.001N$ standard solutions in the black lacquered reservoirs of the piston burets change by about 2% within the first 3 days, but then remain constant for a period of at least 2 weeks. Coulometric iodine addition (25) is also possible.

2. Only 250 μl alkaline absorbent solution is sufficient for the complete absorption of a few micrograms of hydrogen sulfide from a hydrogen stream of 10 ml/min; thus the determination can be carried out in a volume of 500 μl. The principle of differential electrolytic potentiometry (28,29) (cf. Sect. 2.7.2.2) with silver electrodes polarized by a very weak direct current (10^{-8} to 10^{-10} A/cm^2) is used for the endpoint determination. The silver electrodes must be cleaned with potassium cyanide solution before each determination. Special silver sulfide electrodes can be used if only very small sulfide concentrations are to be determined (33).

The titration curve is recorded with a vacuum tube voltmeter with high input impedance and zero suppression (0–200 mV range) or, in using a motor-driven piston buret, with a suitable millivolt recorder.

A sharp, nicely reproducible endpoint is obtained if the electrodes are connected to a 12-V emf source through a 1000 Mohm resistance (cf Fig. 35).

The amount of sulfide deposited on the anode by the polarization current is far below the amount determined and has no effect on the accuracy of the procedure.

A formation of Ag_2S by reaction of the sulfide ions of the solution with the silver electrodes, not electrolytically caused, has been observed only with large electrode areas and very small sulfide ion concentrations.[39]

With an electrode area of <2 mm^2 and a sulfide concentration

---

[39] With a S^{2-} concentration of 0.5 μg S^{2-}/ml and a total electrode area of about 10 mm^2, about 12% of the sulfide ions present in the solution were bound after 30 min.

of >5 μg S^{2-}/ml no exchange can be detected within 30 min. The smallest possible contact times between solution and electrodes will be obtained if the hydrogen sulfide is first bound in a buffered zinc solution and the precipitated ZnS is dissolved in NaOH–NH$_4$OH–EDTA solution shortly before the beginning of the titration. Since the sulfide ions will react with the silver only after addition of the EDTA solution, a reaction can occur only during the titration, which does not last longer than 10 min.

Figure 43 shows a titration curve obtained in the determination of 8.0 μg S^{2-}/ml with a 0.005N silver nitrate solution. The relative standard deviation of the procedure in the range of 3.2–32 μg S^{2-} /0.5 ml is $s_{56} = \pm0.7\%$. In this region the AgNO$_3$ solution consumption is strictly proportional to the amount of sulfide introduced.

Incompletely rinsed cyanide (electrode cleaning) will be titrated after the sulfide determination and will give a second endpoint without affecting the accuracy of the sulfide determination. The procedure is suitable for determination of >2 μg sulfur; it is used for the determination of hydrogen via H$_2$S (cf. Sect. 2.2.2.3).

3. After absorption of the hydrogen sulfide in 0.1 ml 1N sodium hydroxide solution, the sulfide can be titrated directly with 0.001M cadmium acetate solution, forming cadmium sulfide (7). Excess

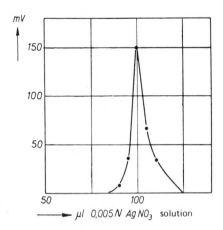

Figure 43. Titration curve for argentometric sulfide determination with differential potentiometric endpoint indication.

Cd^{2+} ions react with dithizone to form intensely red cadmium dithizonate. The recognition of the endpoint is improved by a two-phase titration in which the dithizone is dissolved not in water, but in a carbon tetrachloride phase having only about one-tenth the volume of the aqueous phase (1 ml). The two phases are thoroughly mixed during the titration by a spiral stirrer (cf. Fig. 63) or a gas stream (hydrogen or nitrogen) (cf. Fig. 45). At the endpoint the colorless CCl_4 phase becomes pink. The standard deviation of the procedure is $s_{25} = \pm 10$ ng S^{2-} for 1 μg S^{2-} in about 1 ml.

4. The direct titration of very small amounts of sulfide with NaOCl solution (30,31) can be indicated only poorly electrometrically. On the other hand, the oxidation of sulfide by NaOBr in alkaline solution according to

$$S^{2-} + 4OBr^- \rightarrow SO_4^{2-} + 4\ Br^-$$

can be indicated biamperometrically (cf. Sect. 2.9.2.2). The most favorable NaOH concentration is $2N$, the polarization potential 150—200 mV. Since the procedure takes place in the absence of oxygen, no oxygen step will appear. Chloride, bromide, iodide, fluoride, and phosphate do not interfere. Ammonia is also oxidized; it must be driven out beforehand by flushing the solution with hydrogen. For less than 100 μg NH_3, 5 min flushing (10 ml/min) is adequate.

The following should be heeded in the procedure:

1. A 0.001M NaOBr solution in the black lacquered reservoir of a micro piston buret will change titer by as much as 2%/day during the first few days, but after about 5 days the change will be less than 1%/day. Only glass lines may be used between buret and determination vessel.

2. Toward the end of the titration the oxidation proceeds very slowly. Consequently there are two alternatives: The NaOBr solution must be added at greater intervals toward the end of the titration until the measuring device shows no more change in current caused by NaOBr consumption, or else the NaOBr solution is added by a motor-driven piston buret at a rate of 5–10 μl/min. The endpoint is referred to an alkaline sodium sulfide standard solution. In the titration with $0.001M$ NaOBr solution the standard deviation for 1 μg S in 0.5 ml solution is $s_{15} = \pm 8$ ng sulfur. For the titrimetric determination of very small amounts of sulfur which result from a gas–evolving procedure or from a hydrogenating decomposition, the third and fourth procedures are more advantageous. The titration with cadmium solution is described in Sect. 2.5.3.2 together with the gas-evolving procedure, and the titration with sodium hypobromite solution is described in Sect. 2.5.3.3 along with the hydrogenating decomposition.

2.5.3. PROCEDURES FOR SULFUR DETERMINATION

2.5.3.1. PROCEDURE OF BELCHER, GOUVERNEUR, CAMPBELL, AND MACDONALD (4) FOR MORE THAN 10 μg SULFUR

2.5.3.1.1. *Principle*

The directly weighed sample (containing 10–30 μg sulfur), wrapped in polyethylene film, is burned in the oxygen flask. The combustion products are absorbed in hydrogen peroxide and the sulfate is titrated with $0.01N$ barium perchlorate solution in the presence of ethanol. Thorin serves as indicator. The relative standard deviation of the procedure is given as $\pm 1.2\%$ for samples that contain no phosphorus. Special procedures are given for samples containing phosphorus and iodine.

2.5.3.1.2. *Equipment*

Combustion flask. Cf. p. 42.

Piece of polyethylene film. Cf. p. 43.

Piece of linen thread. Cf. p. 43.

Titration apparatus. Cf. p. 24.

2.5.3.1.3. *Reagents*

Barium perchlorate solution. 3.4 g anhydrous barium perchlorate is dissolved in 200 ml distilled water and 800 ml ethanol (for spectroscopic purposes). The pH of the solution is brought to about 3.5 (glass electrode) with perchloric acid. The solution is standardized against $0.005M$ sulfuric acid as follows: 100.0 μl $0.005M$ sulfuric acid, 100 μl distilled water, 2 ml ethanol, 25 μl Thorin solution, and 15 μl 2.5% perchloric acid are introduced into a 10-ml titration flask. The solution is then titrated with stirring (magnetic stirrer) to the first persistent pink.

The titration is carried out with a white background and a daylight lamp. The titer is calculated from the amount used.

Ethanol. Anhydrous, for spectroscopic purposes.

2.5 volume % perchloric acid. Prepared from 70% perchloric acid a.r. and distilled water. The solution is kept in a ⊽ glass pipet bottle with a 15–μl pipet.

0.1% aqueous Thorin solution. Kept in a polyethylene pipet bottle with a 15–μl pipet.

0.002% methylene blue solution. Kept in a pipet bottle with a 15–μl pipet.

30% hydrogen peroxide a.r.

Magnesium oxide a.r.

2.5.3.1.4. *Decomposition Procedure*

Between 30 and 100 μg sample is weighed in a platinum boat. The sample is transferred to the center of a piece of polyethylene film (15 x 15 mm, dark background) and the boat is reweighed. Now the corners of the sheet are folded over the sample by means of two forceps with platinum tips. The packet is folded about the end of a linen thread which serves as fuse and, by means of two platinum-tipped forceps, wedged into the platinum gauze of the sample holder, with the fuse in the direction of the platinum wire support.

The cleaned combustion flask is loaded with enough hydrogen peroxide to permit wetting the entire wall of the flask. The flask is then filled with oxygen through a glass or polyethylene capillary reaching to the lower part of the flask (30 sec). After it has been ascertained that the sample will be in the exact center of the flask upon introduction of the sample holder, the support stopper is wet with a drop of water, the thread is ignited by an alcohol flame, and the sample holder is quickly brought into the decomposition flask, which is oriented with its opening downward. The flask is left for 90 min protected from dust in a desiccator. The edge of the joint of the flask is then wiped with filter paper moistened with alcohol and the stopper of the sample holder is raised a few millimeters out of the joint. About 40 μl water is sprayed between the stopper and the joint with an injection syringe, the loosened stopper is turned a few times without any water being lost, 1 ml ethanol is then added in portions from an injection syringe, the stopper is removed further, and the stopper and platinum parts are rinsed first with ethanol and then once more with 3 drops of water. The flask is again closed and left in a desiccator for 30 min for the rinse solution to collect. It is then opened and the sample holder and walls are again rinsed with 1 ml ethanol.

2.5.3.1.5. *Titration Procedure*

2.5.3.1.5.1. *In the Absence of Phosphorus and Iodine.* A stirring bar is put into the flask, the flask is clamped into the magnetic stirrer, the base of which is covered with white filter paper, and 25 μl Thorin solution, 15 μl methylene blue solution, and 15 μl perchloric acid solution (2.5%) are added to the solution. With the stirrer running, 0.01M barium perchlorate solution is added from a 500-μl piston buret, only 0.2-μl portions being added near the end of the titration. A daylight lamp is used for observation. Shortly before the end of the titration, the flask is rotated to collect solution spattered on the upper part of the wall.

The endpoint of the titration is attained when the color of the titrated solution corresponds to the color of a standard solution obtained as follows: 25 μl Thorin solution, 15 μl methylene blue solution, and 15 μl 2.5% perchloric acid solution are added to 0.2 ml water and 2 ml ethanol in a titration flask. To this is added just enough 0.01M barium perchlorate solution to color the solution persistent pink (usually 1.1–1.2 μl). The color is not stable, so the standard solution must be freshly prepared for each series of analyses.

To determine the blank consumption, which should normally be about 0.5 μl, the decomposition and determination are carried out analogously without sample.

2.5.3.1.5.2. *In the Presence of Iodine.* Between the first rinsing of the flask with 1 ml ethanol and the waiting period of 30 min (cf. Sect. 2.5.3.1.4), the solution is taken to dryness by heating carefully while passing in a slow stream of oxygen, thus volatilizing the interfering iodine. After the heating, the flask wall is again rinsed with 1 ml ethanol and, after a 30 min wait, the procedure is continued as described in Sect. 2.5.3.1.5.1.

2.5.3.1.5.3. *In the Presence of Phosphorus.* Before carrying out the titration (cf. Sect. 2.5.3.1.5.1), 0.3–0.5 mg magnesium oxide is added to the solution and the mixture is heated to boiling for about 5 min with occasional shaking, allowed to cool for 30 min, and filtered through a thin layer of filter paper pulp with the filtration device shown in Fig. 44. The solution is transferred to the filter with the pipet shown in Fig. 9. The flask is rinsed out with 500 μl ethanol, the vacuum is then interrupted, 50 μl water is put onto the filter pulp, and suction is resumed; flask, pipet, and filter are again washed with 500 μl ethanol, suction is stopped again, another 50 μl water is put onto the filter, suction is resumed, and a third rinse is given, now with 1 ml ethanol. Then 100 μl 2.5% perchloric acid and the indicator solutions are added and the titration is performed as described in Sect. 2.5.3.1.5.1.

The sulfur determination proceeds without interference with samples containing nitrogen and chlorine. Interference by fluorine is prevented to a certain extent by complexation with borate derived from the borosilicate glass, but in the case of an F/S ratio of 21:1 too high sulfur contents were obtained.

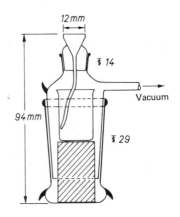

Figure 44. Filtration apparatus after Belcher.

2.5.3.2. PROCEDURE FOR MORE THAN 1 μg SULFUR (7)

2.5.3.2.1. *Principle*

The sample is decomposed on a paper support in the oxygen flask after solution partition (cf. Sect. 1.2.4.1.4). The sulfate is converted by a mixture of hydriodic acid and hypophosphorous acid to hydrogen sulfide, which is determined directly titrimetrically with cadmium solution in the sodium hydroxide absorbent solution.

2.5.3.2.2. *Equipment*

Combustion flask. Cf. Fig. 27, p. 45.

Teflon stopper with electrodes. Cf. Fig. 27, p. 45.

Induction coil (about 10 kV). For example, automotive ignition coil with interrupter.

Hydrogen sulfide evolving apparatus. Details of the borosilicate glass apparatus are shown in Fig. 45. A new apparatus should be cleaned as follows: Concentrated hydrochloric acid is heated in the flask, without cooling the reflux condenser, and is thus distilled through the apparatus for 15 min. The polyethylene capillary *h* should not be in the absorption vessel at this time. Steam is then passed through for about 15 min.

2.5.3.2.3. *Reagents*

Digesting acid. Two parts fuming HNO_3 a.r. (sp. gr. 1.52, $<0.0005\%$ $SO_4{}^{2-}$) and one part $HClO_4$, 70% a.r. ($<0.0005\%$ $SO_4{}^{2-}$). The mixture is kept protected from dust in a ground-joint quartz bottle.

0.1% aqueous calcium nitrate solution. $CaCO_3$ a.r. ($<0.0005\%$ $SO_4{}^{2-}$) is dissolved in a slight excess of HNO_3 and the solution diluted with double-distilled water; it is kept in a polyethylene pipet bottle with a 100–μl pipet in a glove box.

Approximately 1 N sodium hydroxide solution. NaOH a.r. ($<0.0005\%$ $SO_4{}^{2-}$) is dissolved in double-distilled water. It is stored protected from dust in a 100-μl polyethylene pipet bottle. Upon addition of dithizone the solution should not become red, but only yellow orange.

Hydrogen peroxide, 3% a.r. ($<0.0005\%$ $SO_4{}^{2-}$) Stored in a 50-μl polyethylene pipet bottle.

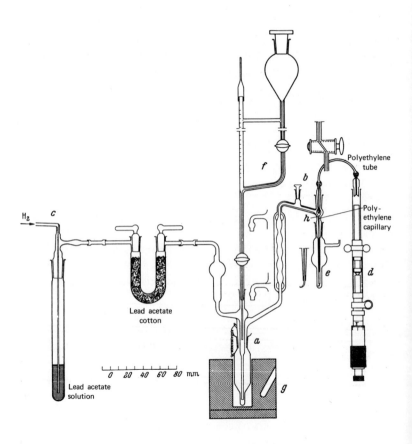

Figure 45. Apparatus for sulfur determination by gas evolution and titrimetric sulfide determination (7). (*a*) Combustion flask with evaporation residue; (*b*) $\overline{\text{S}}$ 5 joint with Teflon stopper; (*c*) bubble counter with lead acetate solution; (*d*) 500-μl micro piston buret; (*e*) absorption vessel; (*f*) reservoir buret for reductant solution; (*g*) metal heating block; (*h*) glass inlet tube, through which passes a polyethylene capillary connected to the piston buret.

Sulfate standard solution. 271.76 mg potassium sulfate a.r. (dried at 110°) is dissolved in some double-distilled water. The solution is brought to volume in a 1000-ml volumetric flask and stored in a polyethylene bottle; 1 ml corresponds to 50.0 μg S. This stock solution should be diluted 1:10.

Reducing solution. 110 ml hydriodic acid a.r. (sp. gr. 1.7), 70 ml concentrated hydrochloric acid a.r., and 10 ml 50% hypophosphorous acid a.r. are heated to a vigorous boil for 30 min in a ground-joint apparatus with a reflux condenser while passing in purified hydrogen. The solution is stored in a brown glass bottle ($\bar{\mp}$ Jena apparatus glass) protected from dust. Because of the danger of sulfate contamination, the bottle should be opened only briefly in replenishing the reservoir of the determination apparatus (cf. Fig. 45).

Acetone a.r. Freshly distilled.

Carbon tetrachloride a.r. For example, E. Merck Co., Darmstadt, Germany, for dithizone determination.

0.05% dithizone–CCl$_4$ solution. Prepared with dithizone a.r.; if protected from light, the solution is stable for about 8 days. Note that solid dithizone, too, is of only limited stability.

Indicator solution. 1 ml 0.5% dithizone–CCl$_4$ solution, 20 ml acetone, and 10 ml CCl$_4$ are mixed before use and stored in the dark. The solution is usable for about 1 day.

0.001 M cadmium chloride solution. 0.1124 g purest cadmium metal is dissolved in as little aqua regia as possible, the excess acid is dispelled, the residue is taken up in 50 ml purest 2 N HCl (obtained by isothermal diffusion), and the solution is diluted to 1000 ml with double-distilled water. The solution will remain titer constant for weeks in a polyethylene bottle. 1 μl corresponds to 32.06 ng S.

Cylinder hydrogen. Cleaned by passing through two drying towers filled with NaOH a.r. pellets, through a bubble counter with 5% lead acetate solution, and through a U-tube filled with lead acetate cotton (polyethylene tubes).

2.5.3.2.4. *Decomposition Procedure in the Oxygen Flask*

In a glove box, 100 μl 1 N sodium hydroxide solution, 100 μl double-distilled water, and 50 μl hydrogen peroxide (3%) are introduced into a combustion flask, and purified oxygen [40] is passed into the flask through a polyethylene capillary. The dissolved sample is on a paper disk which is clamped in the platinum loop of the electrode holder shown in Fig. 27, b (cf. Sec. 1.3.1).

The combustion flask is closed tightly with the Teflon stopper of the electrode holder and the sample is ignited by sparkover. After an absorption time of at least 30 min, during which the absorbent solution is spread over the walls a number of times, the electrodes and joints are rinsed with 1–1.5 ml double-distilled water (rubber bulb pipet). The solution is evaporated at 120°, protected from dust, either in the inclined flask in a drying oven or, better, in the device described on page 48; sulfur is then determined by the gas evolving procedure (cf. Sect. 2.5.3.2.6).

2.5.3.2.5. Decomposition Procedure with $HClO_4$–HNO_3

This decomposition procedure is applicable only to sulfur compounds with SO_3H groups. 100 μl calcium nitrate solution and 200 μl $HClO_4$–HNO_3 mixture are added to the sample in the digestion flask and the mixture taken to dryness at not over 210° (heating block, dust protection).

2.5.3.2.6. Gas-Evolving Procedure

The combustion flask with the evaporation residue is connected to the completely dry apparatus (cf. Fig. 45). (The joint is sealed with a drop of sirupy phosphoric acid a.r.) The apparatus is rinsed several times through opening b with double-distilled water and with pure acetone and is subsequently dried about 5 min with an air stream cleaned and dried by NaOH. The air is then displaced with hydrogen from c, which is adjusted after about 3 min to a flow of 25–30 bubbles/min. The 500-μl piston buret d is filled with cadmium solution, during which the syringe, polyethylene tube connection, and polyethylene capillary must be free of air bubbles. The absorption vessel e is provided with 0.1 ml 1 N sodium hydroxide solution and connected to the apparatus. After addition of 2 ml reducing solution from the reservoir buret f, opening b is closed with a Teflon stopper and the metal block g is heated to about 300°. Boiling is continued 20 min more at this temperature. Before

---

[40] Cylinder oxygen is used which is passed through two drying towers filled with NaOH pellets and one filled with quartz wool.

titrating the hydrogen sulfide absorbed in the sodium hydroxide, the interior of the inlet tube h is rinsed twice (rubber bulb pipet) with 0.5 ml double-distilled water from b, 0.1 ml indicator solution is added, this is rinsed with 0.1 ml CCl_4, and b is closed. The hydrogen stream is adjusted to intermix the two phases vigorously. Titration with the cadmium chloride solution is continued until the originally colorless organic phase just becomes pink. For better observation of the change, the hydrogen flow must occasionally be interrupted. Evaporated carbon tetrachloride can be replenished through opening b.

The blank value and the indicator error are determined by the same procedure. The cadmium standard solution must be standardized from time to time against sulfate standard solution. New apparatus and decomposition flasks are boiled 1–2 hr with reducing solution before use. If the apparatus is not used continually, the first determination usually yields greater variations.

The ubiquity of sulfur necessitates extremely clean work. To control the blank values, it is necessary to work with cleanest reagents and equipment in dust-free and sulfur-free laboratory air. Continual blank value checks are indispensable.

2.5.3.3. PROCEDURE FOR LESS THAN 1 μg SULFUR (8)

2.5.3.3.1. *Principle*

The sample (2–10 μg) is introduced by solution partition and decomposed by hydrogenation. The hydrogen sulfide formed is absorbed in 2 N sodium hydroxide solution and the sulfide titrated with 0.005 M sodium hypobromite solution with biamperometric endpoint indication.

2.5.3.3.2. *Decomposition and Determination Apparatus (cf. Fig. 46)*

The apparatus described in Sect. 2.4.2.3.2 for the determination of nitrogen can be adapted by the following changes: The catalyst tube (B_2) is filled half with pieces of platinum wire (1–2 mm length, 0.3 mm diameter) and half with platinum-pumice (cf. following section). The absorption vessel (B_3) with its two fused-in platinum electrodes (7) is described in Sect. 2.2.2.3.2. It is connected by a polyethylene tube (5) (1.5 mm i.d.) to a 1-ml microburet (Teflon plug and reservoir) containing 2 N sodium hydroxide solution and by a thin glass tube to a motor-driven 500-μl piston buret containing 0.005 M sodium hypobromite solution. The glass body of the piston buret is lacquered black. The glass tube (1.5 mm i.d.) is connected to the buret and the capillary joint (6) of the absorption vessel by pieces of polyethylene tubing (glass–glass contact). The platinum electrodes should be degreased with acetone; they are connected as cathodes against an auxiliary anode in dilute sulfuric acid (at 4–5 V) until a uniform gas evolution appears on their surfaces. For the biamperometric endpoint indication, a polarization emf of 150 mV (B_4) is applied between the two electrodes. The current curve can be recorded over a 50-Mohm resistance by a millivolt recorder (0–200 mV range) with zero suppression or directly by a microampere recorder.

2.5.3.3.3.   *Reagents*

Approximately 0.01 M sodium sulfide solution. Must be kept protected from light and continually vented with oxygen-free hydrogen. Its exact sulfide content must be determined iodometrically daily. The solution is used to standardize the NaOBr solution (cf. Sect. 2.5.3.3.4).

Sodium hypobromite solution, ca. 0.02 M. 1 g sodium hydroxide a.r. and 23 g sodium carbonate monohydrate are dissolved in about 200 ml double-distilled water in a 1000-ml volumetric flask, and 0.5 ml bromine is slowly added with thorough mixing. The solution is diluted to volume with double-distilled water and kept in a glass stoppered bottle of brown borosilicate glass in a refrigerator. Its oxidation value decreases by about 3% within 6 weeks. For the titration a solution diluted to 0.005 M is used which must be standardized before use against the sulfide standard solution as described in Sect. 2.5.3.3.4. The solutions are kept in the dark in glass stoppered bottles of borosilicate glass.

Approximately 2 N sodium hydroxide solution. Prepared by solution of sodium hydroxide pellets in double-distilled water. The solution is kept in a glass bottle, not a polyethylene bottle.

Pieces of platinum wire. Platinum wire of 0.3 mm diameter is cut into 1–2 mm lengths, boiled a few minutes in nitric acid (1:1), thoroughly washed with double-distilled water, ignited, and stored in a glass–stoppered bottle.

Platinum–pumice. Granulated pumice (0.5–1 mm diameter) is saturated with 10% hexachloroplatinic (IV) acid solution a.r., dried, and ignited at 800–900°. It is stored in a glass-stoppered bottle.

2.5.3.3.4. *Determination Procedure*

Before the first measurement, oxygen-free hydrogen is passed through the apparatus for several hours, and the pyrolysis chamber (B_1) (Fig. 46) and the catalyst tube (B_2) are heated to 800–1000°.

The hydrogen flow is adjusted to 10 ml/min; during a measuring period it is not turned off. The syringe body of the buret and the tube to the absorption vessel should be rinsed several times with fresh NaOBr solution from the buret reservoir before each determination.

The catalyst furnace *(3)* is at 900°. After the sample has been inserted into the heating coil of the pyrolysis chamber (B_1), 500 μl 2 N sodium hydroxide solution is introduced into the absorption vessel. The apparatus is rinsed free of oxygen for 10 min. The heating coil is quickly heated to about 900° and this temperature is maintained for 10 min.[41] Afterwards the gas inlet capillary of the absorption vessel is rinsed through *4* with 0.2 ml "oxygen-free" water and the NaOBr solution is added at a rate of 5–10 μl/ min. The beginning of titration is marked on the recorder. The titration endpoint is shown by the current–time curve.

The sulfur content can be determined directly from the consumption of sodium hypobromite solution, standardized against sodium sulfide solution, with the blank consumption subtracted. More exact results, however, will be obtained by calibrating the sodium hypobromite solution with a reference substance, e.g., sulfonal, before and after a series of about 5 determinations.

To standardize the sodium hypobromite solution with sulfide standard solution, 5 or 10 μl accurately standardized (about 0.01 M) sodium sulfide solution is introduced with a 100-μl piston buret into the absorption vessel containing 500 μl 2 N sodium hydroxide solution. The solution is stirred with a steady hydrogen stream of

---

[41] 15 min for samples containing nitrogen.

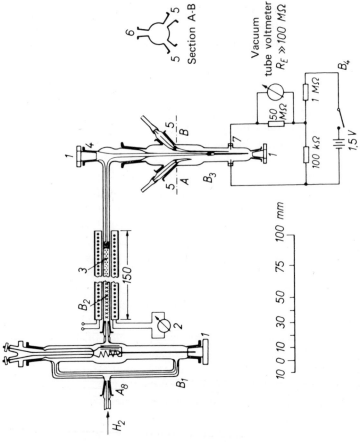

Figure 46. Apparatus for sulfur determination with 2–10 μg sample. A_8, connection to apparatus for hydrogen generation and purification (cf. Fig. 41); B_1, pyrolysis section; B_2, catalyst tube with platinum bed; B_3, absorption and determination vessel; B_4, circuit for biamperometric endpoint determination; *1*, Teflon stopper; *2*, thermocouple; *3*, tube furnace (soldering iron heating element); *4*, $\overline{\underline{\text{S}}}$ 5 joint; *5*, polyethylene tube connections to a 1-ml microburet and a 500-μl piston buret; *6*, connection to a rinsing system (cf. Fig. 42); *7*, platinum electrodes.

10 ml/min and, as described above, titrated with the sodium hypo-bromite solution.

The procedure can be applied to samples containing nitrogen, chlorine, bromine, and iodine. Sulfur compounds containing fluorine and phosphorus have not yet been examined.

2.5.4. REFERENCES

1. W. Merz, Z. Anal. Chem., **207**, 424 (1965).
2. R. Belcher, R. L. Bhasin, R. A. Shah, and T. S. West, J. Chem. Soc., **1958**, 4054.
3. D. C. White, Mikrochim. Acta, **1959**, 254; **1962**, 807.
4. R. Belcher, P. Gouverneur, A. D. Campbell, and A. M. G. Mac-donald, J. Chem. Soc., **1962**, 3033.
5. J. E. Fildes and W. J. Kirsten, Microchem. J., **9**, 411 (1965).
6. W. J. Kirsten, Microchem. J., **7**, 34 (1963).
7. G. Tölg, Z. Anal. Chem., **194**, 20 (1963).
8. G. Tölg, Habilitationsschrift, Mainz, 1965.
9. H. ter Meulen, Rec. Trav. Chim., **41**, 112 (1922); **53**, 121 (1934).
10. E. Wiesenberger, Mikrochim. Acta, **1941**, 73.
11. J. Irmescu and E. Chirnoaga, Z. Anal. Chem., **128**, 71 (1947).
12. H. Stratmann, Mikrochim. Acta, **1956**, 1031.
13. W. J. Kirsten, Z. Anal. Chem., **181**, 1 (1961).
14. W. Geilmann and G. Tölg, Glastech. Ber., **33**, 332 (1960).
15. L. Gustafsson, Talanta, **4**, 227, 236 (1960).
16. J. S. Fritz and M. Q. Freeland, Anal. Chem., **26**, 1953 (1954).
17. J. S. Fritz, S. S. Yamamura, and M. J. Richard, Anal. Chem., **29**, 158 (1957).
18. M. Le Peinter and J. Richard, Chim. Anal., **39**, 331 (1957).
19. E. Pella, Mikrochim. Acta, **1961**, 472.
20. H. Soep and P. Demoen, Microchem. J., **4**, 77 (1960).
21. G. Beilstein, Dissertation, Mainz, 1964.
22. E. Michelski and A. Wtorkowska, Chem. Anal. (Warsaw), **7**, 783 (1962).
23. Z. Kurzawa, Chem. Anal. (Warsaw), **5**, 551, 567 (1960).
24. M. Wronski, Z. Anal. Chem., **180**, 185 (1961).
25. M. Pribyl and Z. Slovak, Mikrochim Acta, **1963**, 1119.
26. C. H. Liu and S. Shen, Anal. Chem., **36**, 1653 (1964).
27. M. W. Tamele, V. C. Irvine, and L. B. Ryland, Anal. Chem., **32**, 1002 (1960).

28. E. Bishop and R. G. Dhaneshwar, *Analyst*, **87**, 845 (1962).
29. G. Schwab and G. Tölg, *Z. Anal. Chem.*, **205**, 29 (1964).
30. O. Bethge, *Anal. Chim. Acta*, **9**, 129 (1953).
31. B. L. Dŭniez and T. Rosenquist, *Anal. Chem.*, **24**, 404 (1952).
32. A. Grünert, K. Ballschmiter, and G. Tölg, *Talanta*, **15**, 451 (1968).
33. A. Grünert and G. Tölg, *Talanta*, in preparation.

2.6. FLUORINE

2.6.1. DECOMPOSITION, GENERAL

Even less than 1 μg fluorine can be determined with sufficient accuracy after combustion of the sample in the oxygen flask (1). Belcher, Leonard, and West (2) used polyethylene film as combustion support for samples of 30–80 μg and observed low values only for substances with over 50% fluorine content, which they attributed to incomplete combustion. The formerly recommended addition of an oxidizing agent (e.g., potassium chlorate) for samples containing CF_3 groups (3) has been proved unnecessary.

The decomposition after Kirsten (4) (cf. Sect. 1.3.1) is more advantageous, especially for samples with very high fluorine content (5), for the samples are exposed to high temperatures for a longer time. Tölg (1) decomposes samples of 2–10 μg on a paper support of 0.2 cm^2 (cf. pp. 43 ff). In using the oxygen flask it is essential that the combustion vessels be of quartz and that they be well steamed before use.

2.6.2. DETERMINATION, GENERAL

The spectrophotometric determination of more than 10 μg fluoride by the lanthanum complex of alizarin fluorine blue (lanthanum alizarin complexan) (6–11) is used by Fernandopulle and Macdonald (5). Sulfate does not substantially interfere with the determination, but with a phosphorus/fluorine ratio over 1 : 2, phosphorus interferes, as does arsenic with an arsenic/fluorine ratio over 1 : 1.

For the most accurate possible determination of less than 10 μg fluoride, a spectrophotometric determination by zirconium Eriochrome Cyanine for 0–2 μg F$^-$ and a zirconium morinate method for 0–0.5 μg F$^-$ are recommended (1). Both methods will now be considered in detail.

2.6.2.1. ZIRCONIUM ERIOCHROME CYANINE METHOD

Of the zirconium lake methods the zirconium Eriochrome Cyanine method (12) is the most reliable (13, 14). Its sensitivity decreases with increasing acidity, while the influence of phosphate, sulfate, aluminum, and other interfering ions is reduced (13). By raising the HCl concentration from 0.7N (12) to 1.2N and largely retaining the remaining conditions of the procedure, the sensitivity is decreased by about 35%; on the other hand, the amount of phosphate or sulfate which would cause an interference corresponding to 0.1 μg F^-/ml is increased from about 6 to 22 μg phosphate and from 30 to 130 μg sulfate. Interference by aluminum is similarly decreased. With a 6-ml determination volume and 50-mm cuvets, a highly reproducible calibration curve for the 0–2.5 μg region can be obtained (cf. Fig. 47).

The standard deviation of the method is $s_{16} = \pm 7$ ng for 1 μg F^- at $22 \pm 1°$.

2.6.2.2. ZIRCONIUM MORINATE METHOD

The amount of fluoride to be determined by spectrophotometric procedures can be decreased if the colored compound can be extracted with a smaller volume of an organic solvent. This always causes a loss of accuracy. Inasmuch as the only strongly colored alizarin complexan fluoride complexes (6, 7) known at present are extractible only under unfavorable conditions, only indirect bleaching procedures are available.

Zirconium morinate has favorable properties. In hydrochloric acid solution it forms a strongly colored, difficultly soluble complex which is colloidal in aqueous solution, but dissolves in higher alcohols, esters, and ketones (15).

Of the large number of solvents examined, methyl isobutyl ketone (MIBK) was found to be especially suited.

The absorption maximum of such a solution is at 415 nm, while

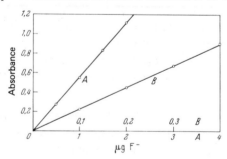

Figure 47. Calibration curves for fluoride determination. *A*, Zr/ECC method, 6-ml determination volume, 50-mm cuvets. *B*, Zr/morin method, 1-ml MIBK, 20-mm semimicro cuvets.

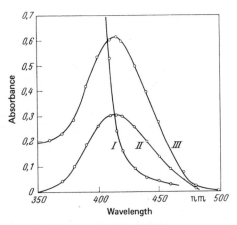

Figure 48. Absorption curves of morin and zirconium morinate in MIBK. Conditions: 9 ml 1 N HCl + 1 ml 0.1% morin–methanol solution with 0 μg (I), 1 μg (II), and 2μg (III) Zr, extracted with 5 ml MIBK and measured in 2-cm cuvets agáints pure MIBK (I) and against I (II and III). Absorption maximum: 415 nm.

pure morin solution absorbs most strongly in the ultraviolet (cf. Fig. 48).

Because of overlapping and mutual effects of the bleaching reaction and the extraction, there are several variables which were examined (1) to locate the optimum bleaching power and reproducibility:

1. Dependence of bleaching on acid concentration of the solution (cf. Fig. 49).

At constant Zr^{4+} and F^- concentrations and equal volumes

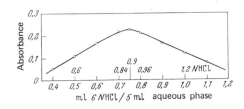

Figure 49. Zirconium morinate bleaching by fluoride as a function of the HCl concentration of the aqueous phase. (Procedure: cf. Sect. 2.6.3.3.4.)

of aqueous and organic phase, the bleaching has a maximum at 0.9 N HCl in the aqueous phase.

2. At constant HCl and F^- concentrations in the aqueous phase, the bleaching shows a strong dependence on the amounts of zirconium and morin added (Fig. 50).

For 0.7 μg F^- there is a bleaching maximum at 3.65 μg Zr per 5 ml aqueous 0.9 N hydrochloric acid phase (0.4 ml 0.0001 M zirconium solution) and about 200 μg morin/ml MIBK. These reagent concentrations must be preserved very exactly in the determination.

3. The time for maximal bleaching amounts to about 60 min. During this time it is necessary to mix the two phases thoroughly with a shaking machine. Since the small volume of the organic phase must in no case change during the shaking, the closure of the vessel must be secure. The previously described digestion flasks with $\overline{\Phi}$ 14.5 joints (cf. Fig. 27, a) are used, closed for shaking with polyethylene stoppers. Glass stoppers (greased or not) are unsuitable.

4. The residual color of the organic phase is constant over a

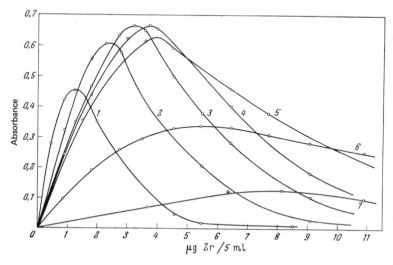

Figure 50. Effect of the amounts of zirconium and morin on the bleaching by 0.7 μg F^-. (Procedure: cf. Section 2.6.3.2.)

period of several hours if evaporation of the solvent, the effect of direct sunlight, and temperature fluctuations over $\pm 1°$ are prevented.

For good reproducibility, sample and standard solutions should be prepared at the same time and under analogous conditions.

Because of the perceptible solubility of water in methyl isobutyl ketone, the organic phase becomes turbid after shaking, when the solution is cooled. It will remain optically clear for a rather long time if the temperature is raised $1–2°$ after the phase separation. It is still better to cool the solution $1–2°$ below room temperature after shaking, briefly shake again, and only then separate the phases. The organic phase is transferred into the cuvet with a suction device (cf. p. 24).

5. Capability of the method: The procedure (Sect. 2.6.3.3.4) yields straight line calibration curves for $0–1$ μg F^- per 5 ml aqueous phase and 1 ml MIBK phase using cuvets with 10-mm optical paths and for $0–0.5$ μg F^- with 20-mm cuvets (cf. Fig. 47).

Reproducibility: Fifteen determinations, each with 250 ng F^-, yielded a relative standard deviation of $\pm 1.1\%$ (measurement temperature: $21 \pm 0.5°$; 20-mm semimicrocuvets; spectrophotometer PMQ II of Zeiss Co. with an additional aperture of 1 mm diameter before the cuvet).

Interferences: 90 μg phosphate or 2 mg sulfate correspond to 1 μg F^-; Fe^{3+} and Al^{3+} may be present only to the order of magnitude of the fluoride.

Phosphate interferes by forming zirconium phosphate, the solubility of which depends strongly on the volume and acidity of the aqueous phase. If the zirconium is first precipitated as phosphate in a small volume, solubility equilibrium in the subsequent greater volume will occur only very slowly. Therefore the two ion species should be brought together only in the final volume of the aqueous phase, i.e., the zirconium solution should be added only as the last reagent when phosphorus is present.

2.6.3. PROCEDURES FOR FLUORINE DETERMINATION

2.6.3.1. PROCEDURE OF BELCHER, MACDONALD, AND CO-WORKERS (2,5) FOR MORE THAN 10 μg FLUORINE

2.6.3.1.1. *Principle*

The sample is wrapped in polyethylene film and burned in the oxygen flask. After absorption of the combustion products in water, the color of the fluoride complex formed with lanthanum alizarin fluorine blue is measured spectrophotometrically.

2.6.3.1.2. *Equipment*

Quartz combustion flask. Cf. p. 42 ff.

Polyethylene film and linen thread. Cf. p. 43.

Spectrophotometer with 40-mm cuvets

100-ml volumetric flasks

2.6.3.1.3. *Reagents.*

$5 \times 10^{-4}M$ *alizarin fluorine blue solution.* 192.6 mg alizarin fluorine blue (alizarin complexan) is first dissolved in as little as possible freshly prepared $2N$ sodium hydroxide solution in a 1000-ml volumetric flask. The solution is diluted with water to about 500 ml, and 50 mg crystallized sodium acetate is added. After this has dissolved, $2N$ hydrochloric acid is added dropwise until the solution is just red (pH 5–6), and then 50 ml acetone and distilled water are added to volume. The solution is mixed thoroughly and filtered through White Band filter paper into a clean, dry, brown glass bottle.

$5 \times 10^{-4} M$ *lanthanum nitrate solution.* 216.5 mg $La(NO_3)_3 \cdot 6H_2O$, dried over phosphorus pentoxide in a desiccator, is dissolved in water in a 1000-ml volumetric flask. The solution is diluted to volume with water.

Acetate buffer solution (pH 4.3). 105 g $CH_3COONa \cdot 3H_2O$ is dissolved in some water in a 1000 ml volumetric flask. After addition of 100 ml glacial acetic acid the solution is diluted to volume with water.

Sodium fluoride a.r. Dried over phosphorus pentoxide.

2.6.3.1.4. *Establishing the Calibration Curve*

Between 20 and 90 μg NaF (corresponding to about 10–40 μg F^-) is weighed on a microgram balance, transferred directly to 100-ml volumetric flasks, and dissolved in some water. Then 10 ml alizarin fluorine blue solution, 2 ml buffer solution, 10 ml lanthanum nitrate solution, and 25 ml acetone are added to each flask and the solution is carefully shaken after the addition of each reagent. The flask is filled to volume with water, shaken again, and left standing at least 90 min protected from direct light. The absorbance of the solution is measured at 620 nm in 40-mm cuvets against a solution containing only the reagents.

For 10–30 μg F$^-$ the points of the calibration curve lie on a straight line passing through the origin. Although the calibration curve is generally reproducible, checking it with a weighed amount of sodium fluoride for each series of measurements is recommended.

2.6.3.1.5. *Decomposition and Determination Procedure*

Between 30 and 80 μg sample is weighed (cf. Sect. 2.5.3.1.4), wrapped in a piece of polyethylene film (15 $\times$ 15 mm), clamped with its fuse in the platinum wire holder, and burned in the combustion flask filled with oxygen and 2–3 ml water. The flask is shaken 5–10 min to absorb the combustion products; the solution is then transferred to a 100-ml volumetric flask and the procedure continued as described in Sect. 2.6.3.1.4.

2.6.3.2. PROCEDURE FOR LESS THAN 2 μg FLUORINE (1)

2.6.3.2.1. *Principle*

The sample, measured out by solution partition, is burned on a paper support in the oxygen flask. The fluoride is absorbed in water and determined spectrophotometrically by the zirconium Eriochrome Cyanine method.

2.6.3.2.2. *Equipment*

Two micro piston burets with reservoirs. Cf. p. 27.

One 5-ml microburet, one 1-ml microburet

Quartz combustion flask. Cf. p. 45.

Teflon stopper with platinum electrodes. Cf. p. 45.

Spark coil. About 10 kV, e.g., automotive ignition coil with interrupter.

Spectrophotometer with 50-mm quartz cuvets

2.6.3.2.3. *Reagents*

Double-distilled water from a quartz still is used in preparing all aqueous solutions.

Fluoride standard solution. 110.53 mg NaF a.r., dried at 110°, and about 2 g solid sodium hydroxide a.r. are dissolved in double-distilled water in a 1000-ml quartz [42] volumetric flask; the flask is filled to volume with double-distilled water. This stock solution with 50.0 μg F$^-$/ml is stable in a polyethylene bottle for several weeks. The working solution with 5.00 μg F$^-$/ml must be freshly prepared each time.

Reagent solution A. 132.5 mg ZrOCl$_2 \cdot$ 8H$_2$O a.r. is dissolved in 200 ml double-distilled water and then 600 ml concentrated hydrochloric acid a.r. (sp. gr. 1.19) is added. After the solution has cooled to 20° it is diluted to 1000 ml with double-distilled water and stored in a quartz vessel.

Reagent solution B. 900 mg Eriochrome Cyanine R a.r. is dissolved in 500 ml double-distilled water. The solution is filtered through filter paper and kept in the dark in a polyethylene flask.

2.6.3.2.4. *Establishing the Calibration Curve*

50–500 μl fluoride standard solution (5 μg F$^-$/ml) is measured with a micro piston buret into 10-ml quartz flasks [cf. Fig. 27, *a*] previously cleaned by boiling with concentrated hydrochloric acid. If glass flasks are used, aluminum may interfere. The volume is then made up to exactly 4.50 ml with double-distilled water from a buret. 1.00 $\pm$ 0.002 ml solution A is added (microburet with polyethylene capillary); the solution is mixed well and brought in a thermostat to a definite temperature between 20 and 24°, which should correspond to the measurement temperature.

With stirring, 500 $\pm$ 2 μl solution *B* is slowly added from a microburet. In measurements made over rather large intervals, combining the two reagent solutions *A* and *B* before addition of the fluoride occasionally yielded rather large fluctuations in the results. After 10 min, the bleaching of the solution is measured at 540 nm in 50-mm quartz cuvets against a fluoride-free, but otherwise analogously prepared, reference solution. The fluoride-containing

---

[42] Use of a glass flask can be expected to produce interference from aluminum.

solution is set to absorbance 0.000 and the absorbance of the
reference solution is measured; its absorbance is constant for about
30 min. The calibration curve is valid for only the stipulated
temperature. It is advisable to check its slope before each series
of measurements. The sensitivity of the method depends on the
quality of the Eriochrome preparation.

2.6.3.2.5. *Decomposition and Determination Procedure*

To burn the sample, 500 ± 2 μl double-distilled water is put into
a quartz combustion flask [cf. Fig. 27, *a*] as absorbent solution and
the flask is filled with purified oxygen (cf. Sect. 2.5.3.2.4), closed
with the electrode holder, and ignited.

The flask is shaken for at least 10 min with a shaking machine to
absorb the combustion products, the polyethylene [43] closure is
opened, and the joint is carefully rinsed with water. The absorbent
solution and rinse water should together amount to exactly 4.50 ml.
The subsequent treatment is described in Sect. 2.6.3.2.4. Spectro-
photometry is carried out against a reference solution obtained in a
similar manner by burning a paper support or by a blank combustion with quartz wool or in a platinum boat.

2.6.3.3. PROCEDURE FOR LESS THAN 0.5 μg FLUORINE (1)

2.6.3.3.1. *Principle*

The sample, measured out by solution partition, is burned on a
paper support in the oxygen flask. The fluoride is absorbed in water
and determined spectrophotometrically by the zirconium morinate
method.

2.6.3.3.2. *Equipment*

See Sect. 2.6.3.2.2. In place of 50-mm quartz cuvets, 20-mm
semimicrocuvets are needed.

2.6.3.3.3. *Reagents*

Fluoride standard solution. Cf. Sect. 2.6.3.2.3. The solution is

---

[43] In this case an electrode holder with a polyethylene stopper, rather than
a Teflon stopper as described on page 44, is used to avoid possible fluoride
blank values due to Teflon.

diluted 1:5 before use with double-distilled water. 1 ml solution corresponds to 1.00 μg F⁻.

0.04% morin–methanol solution. 100.0 mg morin a.r. is dissolved in 250 ml freshly distilled methanol a.r. The solution is stored in a brown glass $ bottle tightly closed with a polyethylene stopper. The bottle may be opened only briefly for removal of solution with a pipet.

0.90 ± 0.01 N hydrochloric acid a.r.

1.00 N hydrochloric acid a.r.

0.0004M zirconium standard solution. 128.9 mg $ZrOCl_2 \cdot 8H_2O$ a.r. is dissolved in 1 N hydrochloric acid and diluted with 1N hydrochloric acid to 1000 ml. The solution is kept in a polyethylene bottle.

Morin–hydrochloric acid solution. 5 ml morin–methanol solution is diluted to volume with 1N HCl in a 50-ml volumetric flask. The solution is prepared fresh daily and vigorously shaken before use.

Methyl isobutyl ketone (MIBK), a.r. Freshly distilled.

2.6.3.3.4. *Establishing the Calibration Curve*

Known amounts of fluoride standard solution, corresponding to 0–0.5 μg F⁻, are measured into 10-ml quartz combustion flasks and taken to dryness in a drying oven at not over 110°. When the flasks have cooled to room temperature, 500 μl 0.9N HCl and 5 ml morin–HCl solution are added, and, after brief shaking, exactly 100 μl zirconium reagent solution is added. Finally 1.00 ±0.001 ml MIBK at 20° is added and the flasks are closed with polyethylene stoppers, forced in tight to prevent any solution creeping out. The flasks are shaken vigorously in a shaking machine for about 60 min and cooled in a thermostat to a temperature 1–2° below the temperature of the measuring chamber. The flasks are then again shaken briefly by hand and replaced in the thermostat until the phases have separated. The organic phase is transferred to a 20-mm semimicrocuvet with a rubber bulb pipet with fine drawn polyethylene tip and the bleaching is measured at 415 nm against an analogously prepared, fluoride-free zirconium–morinate–MIBK solution. The two solutions must have attained the same temperature.

2.6.3.3.5. *Decomposition and Determination Procedure*

In the morin method the combustion products are absorbed in 500 μl 0.9N hydrochloric acid, introduced into the combustion flask before filling it with oxygen. After combustion and absorption, the polyethylene stopper with the electrodes is carefully lifted from the joint, and stopper and joint are rinsed with exactly 5.00 ml morin–HCl solution. Then 100 μl zirconium solution and finally 1.00 ml MIBK are added. Before shaking, the flask is again closed with the electrode holder. The further treatment is analogous to the procedure for establishing the calibration curve. To prepare the reference solution, a combustion is carried out in a similar manner, but without sample.

2.6.4.   REFERENCES

1. G. Tölg, *Z. Anal. Chem.*, **194**, 20 (1963).
2. R. Belcher, M. A. Leonard, and T. S. West, *J. Chem. Soc.*, **1959**, 3577.
3. R. Belcher, M. A. Leonard, and T. S. West, *J. Chem. Soc.*, **1958**, 2390.
4. W. J. Kirsten, *Microchem. J.*, **7**, 34 (1963).
5. M. E. Fernandopulle and A. M. G. Macdonald, *Microchem. J.*, **11**, 41 (1966).
6. R. Belcher and T. S. West, *Talanta,* **8**, 853 (1961).
7. R. Belcher and T. S. West, *Talanta,* **8**, 863 (1961).
8. F. J. Frere, *Anal. Chem.*, **33**, 644 (1961).
9. R. Greenhalgh and J. P. Riley, *Anal. Chim. Acta,* **25**, 179 (1961).
10. F. H. Cox, *Pharm. Weekbl.,* **99**, 801 (1964).
11. M. Buck, *Z. Anal. Chem.*, **193**, 101 (1962).
12. S. Megregian, *Anal. Chem.*, **26**, 1161 (1954).
13. L. L. Thatcher, *Anal. Chem.*, **29**, 1709 (1957).
14. R. Valach, *Talanta,* **9**, 341 (1962).
15. E. B. Sandell, *Colorimetric Determination of Traces of Metals,* 3rd ed., Interscience, New York, 1959.

2.7. CHLORINE

2.7.1. DECOMPOSITION, GENERAL

In the determination of chlorine after decomposition in the oxygen flask, blank value fluctuations of about ± 50 ng Cl^- are obtained even with extremely careful work (1). The determination of about 1 μg chlorine with a relative standard deviation under 2% is possible only after burning the sample in an oxygen stream or after a hydrogenating decomposition in an apparatus completely sealed against the laboratory air (1,2). In comparison, over 5 μg chlorine can be determined very simply with a relative standard deviation of about 1.6% after combustion of the sample in an oxygen flask (3). Inasmuch as the chlorine blank value of the paper, even after special pretreatment, lies between 0.5 and 0.05 μg/0.2 cm² (1), depending on the kind of paper, paper is not a suitable combustion support. Using polyethylene film, one obtains blank values averaging 0.09 μg Cl^- (3). There is no blank value information for the Kirsten decomposition method (4) (cf. Sect. 1.3.1), but we may assume that the very smallest blank value variations could occur with it.

If the sample is not weighed directly, but measured out by a solution partition procedure (cf. Sect. 1.2.4.4), then the technique described on page 45 ff, in which the sample is burned on quartz wool inside an electrically heated Pt–Rh wire spiral, is also good. Nevertheless, the ubiquity of chloride requires extremely clean work. To attain as small blank value fluctuations as possible, it is advisable to conduct all preparations in a glove box in filtered "chloride-free" air. All equipment and reagents should be specially cleaned. According to measurements of List and Tölg (2), for example, in touching a glass object with fingers one transfers about 0.1–0.25 μg Cl^-/cm² onto its surface. After hand washing, the amount of chloride transferred is decreased to about 0.03 μg/cm². A smooth glass surface takes up between 0.01 and 0.2 μg/cm²/hr from the air, depending on local conditions.

2.7.2. DETERMINATION, GENERAL

Of the numerous methods for determination of small amounts of chloride (1), only few are sufficiently accurate.

Mercurimetric titration with diphenylcarbazone as indicator is suitable for over 5 μg chloride (3,5). The standard deviation of the procedure, however, increases rapidly with decreasing amounts of chloride; for 1 μg Cl^-/ml it amounts to about ±40 ng Cl^-. Photometric titration offers no substantial improvement (1). Error results predominantly from blank value fluctuations (reagents and pH adjustment) and from the lack of sharpness of the endpoint. The operations necessary for measuring reagents make determination in a glove box difficult.

Argentometric titration procedures are sensitive enough for a volume of 1–2 ml only if the solubility product of AgCl is reduced by the presence of organic, water-soluble solvents such as alcohols, acetone, dioxane, tetrahydrofurane, and acetic acid and if the endpoint is indicated potentiometrically or amperometrically. In aqueous solution only nullpoint potentiometry (6) has comparable sensitivity, but the method is less suited to small volumes.

In general, the use of glass electrodes or reference electrodes with salt bridges for electrometric endpoint indication cannot be recommended over differential potentiometric procedures, which require only metal electrodes, for very small amounts of chlorine. Differential measuring systems are simpler, and smaller blank values occur. Whereas the usual indication with bimetal electrode systems (7,8) fails to work in organic solvent systems with low chloride concentrations (1,9), even chloride concentrations between 0.2 and 5 μg/ml can be determined very accurately with Ag or AgCl electrodes polarized with a very weak direct current (10). This bipotentiometric titration (differential electrolytic potentiometry) yields especially small blank value fluctuations in glacial acetic acid and makes possible even the determination of 0.5 μg chloride with a relative standard deviation of $\pm1\%$ (1).

2.7.2.1. MERCURIMETRIC TITRATION WITH VISUAL ENDPOINT INDICATION (3)

For the range of 5–30 μg chloride, titration is best carried out in a volume of 2 ml in the presence of 80% ethanol at pH 3.5 with 0.02N mercury(II) nitrate solution. The mercuric nitrate solution is calibrated against NaCl. A mixture of 0.5% diphenylcarbazone and bromphenol blue serves as indicator; it serves for adjusting the pH value and for better recognition of the endpoint.

The ethanol content of the solution may vary between 70 and 87%. In calibrating the mercury(II) nitrate solution against NaCl standard solution, the titer can be reproduced with a relative standard deviation of ±0.2%.

2.7.2.2. ARGENTOMETRIC TITRATION WITH BIPOTENTIOMETRIC ENDPOINT INDICATION (1)

In differential electrolytic potentiometry (10) two equivalent AgCl or Ag indicator electrodes are polarized with a very weak direct current (about 10^{-8} to 10^{-9} A/cm² at the electrodes) (Fig. 51), causing the chloride ion concentration in the immediate vicinity of the anode to be smaller, and in the vicinity of the cathode greater, than in the solution; the potential of the anode in the titration therefore leads and the potential of the cathode lags in relation to the Nernst titration curve for a currentless electrode. If the two electrodes act identically, the potential difference at each point yields the first derivative of the Nernst curve. Since this has a point of inflection at the equivalence point, a maximum occurs here in the differential curve (Fig. 52).

Suitable silver chloride electrodes are obtained by electrolytically silvering and then chlorinating platinum wires fused in glass. In the titration in glacial acetic acid the potential is quickly set.

With increasing current density the endpoint becomes sharper (Fig. 52). The position of the maximum and the endpotential remain unchanged with current densities of 3×10^{-8} to 3×10^{-10} A/cm². The most favorable current density was found to be $5 \times$

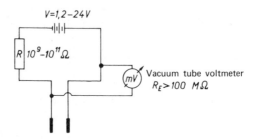

Figure 51. Circuit for bipotentiometric titration after Bishop and Dhaneshwar.

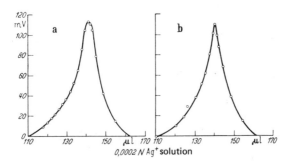

Figure 52. Titration curves: 1 μg Cl$^-$/2 ml glacial acetic acid. *a, R* = 10^{11} ohm, *U* = 1.5 V; current density at the electrodes: 3.75 · 10^{-10} A · cm^{-2}. *b, R* = 10^9 ohm, *U* = 120 V; current density: 3 · 10^{-8} A · cm^{-2}.

10^{-9} A/cm^2 (1), but it is advisable to seek the best polarization conditions for each electrode pair, which may be somewhat different from case to case, by changing the resistance or the potential. In the range of 0.2–5 μg per 2 ml glacial acetic acid, the consumption of silver nitrate is proportional to the amount of chloride. With 0.0002N silver nitrate–glacial acetic acid solution and titration steps of 0.5 μl, the potential is set within a few seconds. The potential change is decreased with increasing water content of the acetic acid (cf. Fig. 53).

A water content of 20% is barely acceptable. Moreover, the method is sensitive to foreign ions; e.g., upon addition of 400 μg sodium acetate (anhydrous) per 2 ml glacial acetic acid only a low maximum appears in the determination of 1 μg chloride. At 1000 μg sodium acetate the electrodes no longer react. To avoid the effect of temperature (cf. Sect. 2.8.2.3), the titration vessel should be thermostated (cf. Fig. 55).

When AgCl electrodes are used, bromide, iodide, and sulfide interfere, since a stepwise titration is not possible (2). With pure silver electrodes, Cl$^-$, Br$^-$, and I$^-$ cannot be determined together very accurately, of course, if they are present to the same order of magnitude. Figure 54 shows a titration curve obtained when 1 μg Cl$^-$ and 1 μg Br$^-$ were titrated in 2 ml glacial acetic acid with 0.0002N silver nitrate solution (cf. Sect. 2.8.2.3).

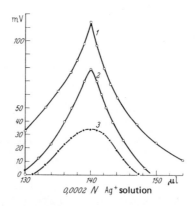

Figure 53. Titration curves: 1 μg Cl$^-$/2 ml acetic acid–water mixture; acetic acid concentration: curve 1—96%; curve 2—80%; curve 3—70%; current density: $5 \cdot 10^{-9}$ A $\cdot$ cm^{-2}.

2.7.3. PROCEDURES FOR CHLORIDE DETERMINATION

2.7.3.1. PROCEDURE OF BELCHER, GOUVERNEUR, AND MAC-DONALD (3) FOR MORE THAN 5 μg CHLORINE

2.7.3.1.1. *Principle*

After decomposition of the weighed sample by combustion in the oxygen flask, the chloride formed is titrated mercurimetrically.

2.7.3.1.2. *Equipment*

Combustion flask. Cf. p. 42.

Polyethylene film. Cf. p. 43.

Linen thread. Cf. p. 43.

Titration apparatus. Cf. p. 24

2.7.3.1.3. *Reagents*

Distilled water. From a glass still.

Mercury(II) nitrate solution. 3 g mercury(II) nitrate a.r. is dissolved in 500 ml 0.01N nitric acid; after 48 hr the solution is filtered

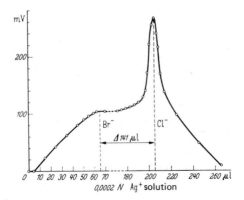

Figure 54. Titration curve: 1 μg Cl⁻/2 ml glacial acetic acid with 1 μg Br⁻; silver electrodes; current density: $5 \cdot 10^{-9}$ A · cm⁻².

(filter paper) into a 100-ml volumetric flask and diluted to volume with distilled water.

Aqueous 0.03% bromphenol blue solution. Stored in a 20-μl polyethylene pipet bottle (cf. p. 22).

Alcoholic 0.5% diphenylcarbazone solution. Stored in a polyethylene pipet bottle with a 20-μl pipet.

Ethanol, purest. For spectroscopic use. Should contain <0.3 μg Cl⁻/ml. Ethanol with a higher chloride content should be redistilled over silver nitrate.

0.05N nitric acid. By dilution of nitric acid a.r.; stored in a polyethylene pipet bottle with a 15–μl pipet.

Sodium chloride, a.r.

Oxygen. The oxygen used to fill the combustion flasks is taken from a steel cylinder, purified through a tube filled with soda asbestos, and passed into the combustion flask through a polyethylene tube drawn out to a capillary. The polyethylene tube should be well rinsed with water. Above all, the part of the tube to be led into the combustion flask must be kept scrupulously clean.

2.7.3.1.4. *Standardization of Mercury(II) Nitrate Solution*

Portions of sodium chloride of 20–40 μg are weighed by difference with an ultramicrobalance into little titration vessels (60 mm length, 15 mm diameter). The wall of each tube is carefully rinsed with 100 μl water from an injection syringe and then 2.0 ml ethanol is added, also from an injection syringe.

For the titration, a flask is placed on the magnetic stirrer and a well rinsed stirring bar is put into the flask with platinum tipped forceps; then 15 μl bromphenol blue solution, about 15 μl 0.05N nitric acid (just enough to color the solution pure yellow), and about 20 μl diphenylcarbazone solution are added. Then the buret tip is dipped into the solution and titration is continued to the first lasting reddish coloring. Just before the end of the titration, the flask is rotated to wash down the chloride on the wall over the solution, and titration is carried out in steps of 0.05 μl mercuric nitrate solution.

It is advisable to vary the indicator concentration somewhat in trial titrations to obtain the most favorable color change for the observer. The relative standard deviation of the titer was found to be less than $\pm0.2\%$.

2.7.3.1.5. *Decomposition and Determination Procedure*

Samples of 30–80 μg containing a maximum of 25 μg chlorine are weighed (cf. Sect. 2.5.3.1.4), wrapped in a piece of polyethylene film with a standard length of linen thread, and clamped in the platinum gauze of the sample holder with the fuse projecting in a line with the platinum suspension wire. The cleaned combustion flask is again rinsed with water so that the entire wall remains wet. The flask is then filled with purified oxygen (cf. Sect. 2.5.3.1.4). After it has been confirmed that the sample is exactly in the center of the flask, the stopper is wet with a drop of water, the thread is ignited at a small alcohol flame, and the sample holder is quickly inserted into the inverted decomposition flask. The latter is then placed for 90 min in an empty desiccator. The rim of the ground joint is then wiped with filter paper moistened with ethanol, the stopper is raised a few millimeters, and 100 μl distilled water is introduced through the joint from a hypodermic syringe. The loosened stopper is rotated several times so that no water is lost. Now 1 ml ethanol is added in portions from a hypodermic syringe while the stopper is withdrawn further. The platinum wire and gauze are rinsed three times with a drop of water from the syringe, each drop being drained onto the vessel wall. Finally the joint is rinsed with 500 μl ethanol; the flask is closed and rotated for 3 min until the entire wall is wet, but not the joint or stopper.

The flask is allowed to drain for 45 min and then the rinsing is repeated with another 1 ml ethanol and 3 drops of water. This is followed by the titration (cf. Sect. 2.7.3.1.4).

Blank combustions are carried out analogously. The blank consumption should not exceed 1.2 μl 0.02N mercuric nitrate solution.

2.7.3.2.   PROCEDURE FOR 0.5–5 μg CHLORINE (1)

2.7.3.2.1.   *Principle*

The sample is weighed directly or measured out by solution partition. Decomposition is by combustion in an oxygen stream. The hydrogen chloride formed is absorbed in glacial acetic acid and determined by argentometric titration with bipotentiometric endpoint indication.

2.7.3.2.2.   *Equipment*

Decomposition and determination apparatus. The apparatus (Figs. 41 and 55) consists essentially of the elements previously described in Sects. 1.3.1 and 2.4.2.3.2: an electrolysis vessel (Fig. 41, A_1) for generation of oxygen,[44] a bubble counter (A_2) filled with water, a U-tube (A_3) with solid sodium hydroxide, a cold trap (A_5) (Dry Ice–methanol), the decomposition vessel (Fig. 55, B_1) with sample holder (B_2) and Pt–Rh heating coil (B_3), and the absorption and determination vessel (C_1).

The path between decomposition vessel and absorption vessel has a very small surface area. The combustion chamber and connecting path can be heated by the Pt–Rh coil. For better heat transfer, the combustion chamber and connecting capillary are wound with 1-mm silver wire. The absorption and determination vessel (C_1) is provided with a cooling jacket so that the absorbent solution may be thermostated. The construction and operation of the high speed dc motor-driven bell stirrer (C_4) in the absorption vessel were described in Sect. 2.2.2.2.2.

The electrodes (0.3 mm diameter, 2 mm effective length) necessary for differential electrolytic potentiometry are in a glass support

---

[44] The center electrode is connected as anode.

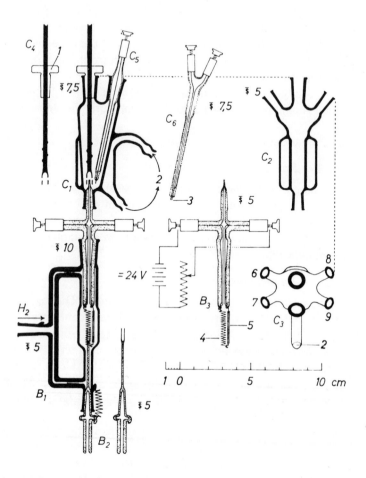

Figure 55. Apparatus for the determination of very small amounts of chlorine and bromine (1,2). B_1, quartz decomposition vessel; B_2, quartz sample holder; B_3, Pyrex heating coil holder; C_1, absorption and titration vessel, side view; C_2, absorption and titration vessel, side view (rotated 90° about its long axis with respect to C_1); C_3, absorption and titration vessel, view from above; C_4, bell stirrer; C_5 and C_6, electrode rod; *1*, Teflon stopper; *2*, water connection to and from thermostat; *3*, platinum electrodes; *4*, heating coil of 0.3 mm Pt–Rh wire; *5*, quartz insulating tube; *6*, $\overline{\overline{\mathbb{S}}}\,5$ joint for connection to 500-μl piston buret; *7*, $\overline{\overline{\mathbb{S}}}\,5$ joint for connection to a 2-ml semimicroburet; *8*, $\overline{\overline{\mathbb{S}}}\,5$ joint for connection to a rinsing system (cf. Fig. 41, B_5); *9*, $\overline{\overline{\mathbb{S}}}\,5$ joint for introduction of standard solutions.

(C_5 and C_6) that is introduced into the titration vessel interchangeably; the electrodes may [in modification of an earlier procedure (1)] be removed for rather long work pauses and stored protected from light in glacial acetic acid. The absorption vessel is provided with four 〒 5 inlet joints (C_3). One is for admission of the titration capillary (polyethylene); it is connected via a capillary joint and polyethylene tube (2 mm diameter) to a 500-μl glass piston buret with micrometer screw or to a motor buret (cf. p. 27). The glass body of the buret should be lacquered black. Glacial acetic acid can be introduced from a 2-ml microburet through the second joint. The buret is closed dust-tight with a tube filled with sodium hydroxide pellets and quartz wool. The third opening is for drawing off the titrated solution and rinsing the vessel by means of the rinsing system (Fig. 41, B_5) described in Sect. 2.4.2.3.2, which largely prevents introduction of chloride into the vessel. Chloride standard solution can be introduced through the fourth opening.

For the electrical circuit (cf. Fig. 51), a 24-V anode battery, a vacuum tube voltmeter (0–500 mV) with input resistance over 100 Mohm and zero suppression or a corresponding millivolt recorder, high ohmic resistances of 10^9, 10^{10}, and 10^{11} ohm[45] and shielded cable are needed.

100–μl displacement buret. Cf. p. 27 ff. For introduction of the chloride standard solution.

2.7.3.2.3. *Reagents*

0.001M chloride standard solution. A known amount, ideally 58.45 mg, of sodium chloride a.r.[46] is dissolved in a 1000-ml quartz volumetric flask [47] in some "chloride-free" water [48] and diluted to volume, and the titer is calculated. Kept in the same flask, the solution will remain titerconstant for several months.

Glacial acetic acid with a chloride content <0.01 μg/ml. Glacial acetic acid a.r. ($<0.0001\%$ Cl^-) with some silver powder added

---

[45] For example, Hi-Meg Resistors of Victoreen Components Div., Cleveland 3, Ohio.

[46] The sodium chloride should be dried for several hours at 180–200° and stored over P_2O_5 before weighing.

[47] The flask should be steamed for at least 30 min before use.

[48] After addition of some NaOH a.r., distilled water is redistilled from a quartz still.

is distilled through a column. The still should be steamed for about 2 hr before the beginning of the distillation. The acid is stored in a previously steamed brown borosilicate glass bottle, protected from dust.

0.0002N silver nitrate–glacial acetic acid solution. 5.00 ml aqueous 0.02N silver nitrate solution, weakly acidified with nitric acid, is diluted to volume with chloride-free glacial acetic acid in a 500-ml quartz volumetric flask.[47] If the solution is stored in a black lacquered polyethylene bottle,[49] the silver ion concentration will hardly diminish perceptibly in the course of months. The same is true for storage of the solution in the black lacquered reservoir of the 500-μl piston buret (cf. p. 27). The silver nitrate solution should be standardized daily against the chloride standard solution (cf. Sect. 2.6.3.2.5).

Methanol, ethanol, and acetone with chloride content <0.01 *ppm.* Upon addition of some solid silver nitrate, the solvents should be distilled through an efficient column and stored in steamed brown borosilicate glass bottles protected from dust. They are used for solution partition and for rinsing equipment.

Silver bath for preparation of silver electrodes. Cf. Sect. 2.7.2.2. 2.7 g silver cyanide and 2.6 g potassium cyanide are dissolved in 100 ml double-distilled water. (Silver cyanide is prepared by mixing equimolar amounts of dissolved silver nitrate and potassium cyanide; the precipitate is filtered off, washed free of nitrate, and dried for 1 hr at 120.°)

2.7.3.2.4. *Apparatus Preparation*

The two platinum wires (*3*) fused in the electrode holder (Fig. 55, C_6) are first cleaned with 30% nitric acid and then carefully ignited. Both electrodes are then connected anodically and cathodically alternately for several minutes in 2N sulfuric acid (6-V battery

---

[49] The polyethylene bottle is cleaned by letting it stand overnight filled with semiconcentrated nitric acid, rinsing it thoroughly with water, and then letting it stand several hours filled with double-distilled water, changing the water several times. About 100 ml 0.0002 *N* silver nitrate–glacial acetic acid solution is then put into the bottle and it is shaken for at least 2 hr in a shaking machine. Then the solution is poured out and the bottle is rinsed twice with 50-ml portions of 0.0002 *N* silver nitrate–glacial acetic acid solution before being filled with the solution to be stored.

and platinum auxiliary electrode). After thorough rinsing with double-distilled water, the electrodes are put into the silver bath in the electrolysis vessel (cf. Fig. 56), and electrolysis is carried out for 1–2 hr with stirring. The platinum electrodes are connected as cathodes against a platinum wire auxiliary electrode (Fig. 56, *3*). With an emf of 6 V, the electrolysis current should be 350–400 μA. (The necessary series resistance R is about 10 kohm.) The silvered electrodes are thoroughly rinsed with water [50] and are

---

[50] The electrodes should remain connected to the current source while being rinsed with water.

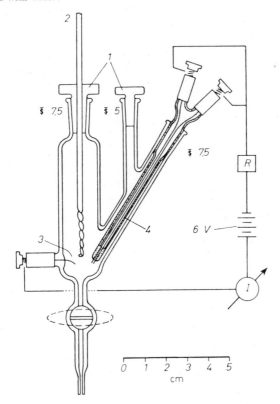

Figure 56. Electrolysis vessel for preparation of Ag, AgCl, or AgBr electrodes for bipotentiometric chloride or bromide determination. *1*, Teflon stopper; *2*, spiral stirrer; *3*, platinum electrode; *4*, electrode rod with platinum electrodes to be silvered.

chlorinated by being connected as anodes against a platinum wire electrode (*3*) for 10 min in 0.5*N* hydrochloric acid (400–500 μA electrolysis current). They are rinsed free of chloride with double-distilled water [50] and are stored in glacial acetic acid protected from direct sunlight. The electrodes react immediately in the titration, but the potential maxima increase during the first few measurements. As a rule, an electrode pair can be used for more than 200 measurements.

Before use, a new piston buret should be etched (room temperature) for a few minutes with a 2–4% hydrofluoric acid solution, thoroughly rinsed several times with distilled water, and steamed for at least 30 min. It is then filled with silver nitrate titrating solution, which is renewed after several hours. After this treatment, the titer of the silver nitrate solution will change only slightly in the first 2 days. The semimicroburet used to measure the glacial acetic acid should be thoroughly cleaned before use with concentrated nitric acid (not with chromosulfuric acid), rinsed several times with double-distilled water, rinsed with glacial acetic acid, and then filled.

To ascertain the most favorable titration conditions, the chloride standard solution is introduced from a 100-μl piston buret (cf. p. 27 ff) into 1 ml glacial acetic acid and titrated with various current densities at the electrodes, different influx rates of silver nitrate solution, and various paper speeds of the recorder.

2.7.3.2.5. *Decomposition and Determination Procedure*

The apparatus is flushed for about 5 min with oxygen (20–50 ml/min, 300 mm water pressure). The oxygen flow should not be interrupted during the determination. With an oxygen flow <20 ml/min there is danger of liquid in the absorption vessel creeping back into the combustion chamber. With the absorption vessel empty, the quartz combustion vessel containing the sample is brought about two-thirds into the Pt–Rh heating coil. After the absorption vessel has been rinsed a number of times with the rinsing system, 1 ml glacial acetic acid is introduced from the semimicro-buret. The bell stirrer is turned on and the sample is burned by rapid heating of the Pt–Rh coil to over 1100°. The coil is kept at this temperature for about 30 sec and is then adjusted to about 900°. The time needed for transfer of the combustion products depends on the quantity of chloride. The oxygen flow rate has no perceptible effect on the transfer time. The time required increases at lower heating temperature. The time necessary for complete transfer of the chlorine must be separately determined for each apparatus by calibration. The time and temperature of heating must then be kept constant for all measurements, for otherwise a displacement of the adsorption–desorption equilibrium between HCl stream and vessel surface will cause high or low results. During the titration neither the stirrer nor the oxygen stream should be turned off. It is necessary to wait until the temperature of the absorbent solution has attained the temperature of the coolant in the absorption vessel jacket, and then silver nitrate solution is added continuously (2–3 μl/sec) from the piston buret. When a potential change is noted on the vacuum tube voltmeter, only 0.5-μl portions of silver nitrate are added, in steps, and the change of direction of the pointer deflection is observed. To test the completeness of HCl transfer, a checking titration is performed after a few minutes.

In using a motor-driven buret, it is best to titrate with a flow rate of about 25 μl/min and a recorder paper speed of about 1.5 m/hr. The paper distance between the beginning of titration and the curve maximum is evaluated, and from this, after calibration with known amounts of chloride, the $AgNO_3$ solution consumption is obtained. Automatic titration reduces the standard deviation of the determination by nearly half. Before the next titration the piston buret is refilled and the vessel is sucked dry and thoroughly rinsed with double-distilled water.

With a new apparatus or after a measurement pause of more than one day, a sample with about 4–5 μg chlorine is first burned and the apparatus is heated until no more chloride can be check titrated in the vessel.

The titer of the $AgNO_3$ solution is tested daily; to do so, 1 ml glacial acetic acid and a known amount of chloride standard solution are introduced from an ultramicroburet (cf. p. 27 ff).

With samples containing bromine and iodine the procedure is analogous except that, in place of AgCl electrodes, pure Ag electrodes are used which are freshly prepared after each determination (cf. Sect. 2.7.3.2.4). The electrolysis time can then be shortened to 15 min. The electrodes should be thoroughly rinsed with double-distilled water after silvering. The electrolysis current should not be interrupted during the rinsing until the electrodes are completely free of cyanide.[51] Freshly prepared electrodes are stored in glacial acetic acid in the dark until used.

2.7.4. REFERENCES

1. G. Schwab and G. Tölg, *Z. Anal. Chem.,* **205**, 29 (1964).
2. W. H. List and G. Tölg, *Z. Anal. Chem.,* **226**, 127 (1967).
3. R. Belcher, P. Gouverneur, and A. M. G. Macdonald, *J. Chem. Soc.,* **1962**, 1938.
4. W. J. Kirsten, *Microchem. J.,* **7**, 34 (1963).
5. J. V. Dŭbský and J. Trtílek, *Mikrochemie (Wien),* **12**, 315 (1953).
6. H. V. Malmstadt and J. G. Winefordner, *Anal. Chim. Acta,* **20**, 283 (1959).
7. K. Schwarz, *Mikrochemie (Wien),* **13**, 1 (1933).
8. B. Cunningham, P. L. Kirk, and S. C. Brooks, *J. Biol. Chem.,* **139**, 11 (1941).
9. E. Bishop and R. G. Dhaneshwar, *Analyst,* **88**, 433 (1963).
10. E. Bishop and R. G. Dhaneshwar, *Analyst,* **87**, 845 (1962).

---

[51] The electrolysis current should now be < 5 μA.

2.8. BROMINE

2.8.1. DECOMPOSITION, GENERAL

Decomposition in the oxygen flask leads to smaller blank value fluctuations in bromine than in chlorine determination, so that in principle even very small amounts of sample can be analyzed. Since, however, compounds with oxidation states other than that of HBr are also formed in oxidizing decomposition, a further reaction must always take place before the determination. Especially for amounts of bromine less than 10 μg, this step readily causes errors; thus a hydrogenating decomposition (1) is preferable for the determination of very small amounts of bromine. This procedure permits determination of bromine with samples of 1–10 μg with a standard deviation of ±13 ng (2).

For over 10 μg bromine, an adaptation of the Schöniger procedure to the microgram range (3) is suitable, while fusion decomposition with sodium or potassium (4) is too complicated and susceptible to interference.

2.8.2. DETERMINATION, GENERAL

Iodometric titration as bromate (5) by the procedure of Belcher, Gawargious, Gouverneur, and Macdonald (3) yields a standard deviation of about ±0.05 μg Br (according to the present author's measurements). The method is recommended for over 10 μg bromine, especially since only iodine is codetermined and chlorine does not interfere.

Fennel and Webb (6) prefer photometric determination with cresol red. They have determined over 2 μg bromine with a standard deviation of ±0.08 μg Br. Chlorine and iodine interfere. The determination is particularly simple, but not very accurate. Both procedures use samples of at least 30 μg.

Of the methods for determination of still smaller amounts of bromine (7–11), argentometric titration with bipotentiometric endpoint indication (10,12) (cf. Sect. 2.7.2.2) is especially advantageous. The procedure permits the determination, with laboratory air excluded, of hydrogen bromide, formed by hydrogenating decomposition, with a standard deviation of about ±5 ng (2). The determination is possible, within limits, in the presence of chloride, iodide, and sulfide.

2.8.2.1. IODOMETRIC DETERMINATION WITH VISUAL ENDPOINT INDICATION (3)

Before the determination the bromine, which is in several oxidation states in the digest, must be oxidized to bromate; sodium hypochlorite solution buffered to pH 6–6.5 has proved good for this. The pH, temperature, and heating time given in the procedure must be adhered to exactly. Organic solvents should not be present, for otherwise the oxidation does not proceed quantitatively. The bromate formed is reacted with iodide in the presence of ammonium molybdate as catalyst and the iodine titrated with 0.01 N thiosulfate solution (Thyodene or starch indicator).

2.8.2.2. COLORIMETRIC DETERMINATION WITH CRESOL RED (6)

If bromine, obtained by oxidation of bromide with bromate in sulfuric acid solution, is allowed to react with dyes of the phenol red type, more strongly colored brominated compounds are obtained, the absorbance of which can be measured photometrically. Bromine determination with cresol red (o-cresolsulfonphthalein), which reacts with two bromine atoms, is more sensitive than the longer known determination with phenol red (13), which reacts with four bromine atoms. Neither hydrochloric nor nitric acid should be present, for they attack the dye. The reaction is strongly temperature dependent; in 1.8 N sulfuric acid at 30° it should be complete after 1 hr. Further oxidation of the dye by excess bromate can be stopped by addition of sodium arsenite. The sensitivity of the method depends strongly on the concentrations of cresol red and sulfuric acid; all instructions of the procedure must therefore be adhered to exactly.

While 15 mg of Na $^+$, K$^+$, or sulfate and 1.5 mg of fluoride, nitrate, or phosphate do not interfere, even 15 μg chloride or iodide, as well as oxidizing or reducing compounds, produce positive errors.

2.8.2.3. ARGENTOMETRIC DETERMINATION WITH BIPOTENTIO-
METRIC ENDPOINT INDICATION (2)

The bromide determination procedure corresponding to that used
for argentometric chloride determination in glacial acetic acid (cf.
Sect. 2.7.2.2) operates with two identical AgBr or pure silver electrodes which are polarized (circuit: Fig. 51) and connected to a
vacuum tube voltmeter with high input resistance or to a millivolt
recorder. The current density necessary to attain the sharpest possible maximum is about 12×10^{-8} A/cm^2.

A change in electrode current density affects the shape of the
curve maximum and also, at greater current densities, the position
of the maximum. Since electrode pairs prepared under identical
conditions may differ in active surface area, the best potential for
a given high ohmic resistance must be determined; it is usually
ca. 6–10 V for a resistance of 1000 Mohm and for platinum wire
electrodes of about 5 mm^2 area (0.5 mm diameter, 3 mm length)
which have been silvered and brominated according to the procedure given in Sect. 2.7.3.2.4.

With proper treatment, an AgBr electrode pair will react properly for about 200 determinations over a period of about 6 weeks.
All ions which, like chloride, iodide, and sulfide, also react to
AgBr electrodes will interfere with the bromide determination.
The interfering ions do not produce individual maxima; only one
maximum occurs, corresponding to the sum of the ions. Because of
its ubiquity (air, reagents, etc.), chloride is the most important
source of interference (cf. Sect. 2.7.1). Therefore all equipment
and reagents must be specially cleaned, stored, and handled to
preclude introduction of chloride (cf. Sect. 2.7.3.2.3).

If bromide is to be determined in the presence of chloride,
iodide, or sulfide, pure silver electrodes must be used to obtain
discrete maxima. The determination is then more lengthy and
somewhat less accurate, however, and each electrode pair can be
used for only a few determinations. The silver platings must afterwards be anodically dissolved in semiconcentrated nitric acid and
the platinum electrodes freshly silvered under constant conditions
(cf. Sect. 2.7.3.2.4). The electrodes must be conditioned (cf.
p. 161) before the actual determination, for otherwise the endpoint cannot be accurately reproduced. After the determination, a

calibration with bromide standard solution is conducted, to which the consumption in the determination is referred. After one or two parallel determinations the electrodes will fail. Note that discrete maxima for the individual halides and sulfide are obtained only if the electrodes are polarized after all the anions to be determined are present in the solution; otherwise the titration curve will proceed as in Fig. 54.

1 μg bromide can be determined without interference in the presence of about 10 μg chloride or iodide or about 5 μg sulfide. With increasing amounts of chloride the error increases, since less distinct maxima occur. Larger amounts of sulfide or iodide poison the electrodes. The chloride maximum follows the bromide maximum, while the sulfide and iodide maxima precede. In both cases the separation between the maxima corresponds to the silver nitrate solution consumption for the amount of bromide. In the presence of iodide or sulfide, bromide and chloride do not yield discrete maxima.

In the determination of 1 μg bromide with silver bromide electrodes, using a motor-driven 500-μl piston buret (cf. p. 27) connected to a millivolt recorder (0–250 mV), the standard deviation was $s_{23} = \pm 4.5$ ng when the addition rate of silver nitrate solution was about 30 μl/min and the recorder paper speed was 2.5 cm/min.

With pure silver electrodes, freshly prepared for each determination, a standard deviation $s_{20} = \pm 20$ ng bromine was obtained in the presence of 3 μg chloride, iodide, or sulfide (measuring range: 50 mV; silver nitrate addition rate: 15 μl/min; paper speed: 0.5 cm/min).

In both cases the position of the endpoint is independent of the stirring speed at a speed >3000 rpm. In the temperature range of 20–50°, a solution temperature change of $\pm 1°$ causes a linear displacement of the endpoint of about $\pm 0.3\%$.

With the form of apparatus found suitable for rapid transfer of the hydrogen bromide from the decomposition into the determination section, the temperature in the determination vessel may vary greatly; it must therefore be thermostated.

As in the chloride determination, there is linearity in the bromide determination between the amount of bromide introduced and silver nitrate consumed at constant determination volume (1 ml).

The dependence of the shape of the titration curve on water content of the glacial acetic acid and on foreign salts is also comparable to the behavior described for the chloride determination.

2.8.3. PROCEDURES FOR BROMINE DETERMINATION

2.8.3.1. IODOMETRIC PROCEDURE FOR MORE THAN 10 μg BROMINE (3)

2.8.3.1.1. *Principle*
After decomposition of the weighed sample (30–80 μg) by combustion in the oxygen flask, the bromine is oxidized to bromate with hypochlorite, the bromate is reacted with potassium iodide, and the liberated iodine is titrated with thiosulfate solution against starch.

2.8.3.1.2. *Equipment*
Combustion flask. Cf. p. 42 ff.

Polyethylene film and linen thread. Cf. p. 43.

Electrically heated aluminum block with thermometer. Cf. Fig. 29. ($\pm 1°$)

Titration apparatus. Cf. p. 24.

2.8.3.1.3. *Reagents*
Ca. 0.01N sodium thiosulfate solution. Contains 100 mg sodium carbonate a.r. per 1000 ml and is prepared with freshly boiled water. The solution is standardized against pure, dried potassium hydrogen iodate as follows: 60–80 μg KH(IO$_3$)$_2$ is weighed into a dried flask. The wall of the flask is rinsed with 2 ml distilled water, 50 μl (2 microdrops) of 2N sulfuric acid and 150 μg (3 drops) potassium iodide solution are added, and the solution is titrated with thiosulfate as described in Sect. 2.8.3.1.4.2.

10% potassium iodide solution. The solution must be prepared fresh daily; it is kept in a brown glass bottle with a pipet.

Ca. 1N sodium hypochlorite solution. Prepared by dilution of commercial "bromine-free" hypochlorite solution. The solution should be kept in a brown glass bottle and freshly prepared after 2–3 weeks.

10% (g/ml) potassium dihydrogen phosphate solution. From potassium dihydrogen phosphate a.r. after Sörensen.

Sodium formate solution. 9.5 ml 90% formic acid a.r. is added dropwise to a solution of 9 g sodium hydroxide a.r. in 20 ml water. The solution is kept in a polyethylene pipet bottle (cf. p. 22); it will keep for about 2 weeks.

3% aqueous ammonium molybdate solution (g/ml). Kept in a polyethylene pipet bottle.

12 N sulfuric acid solution. Kept in a glass pipet bottle.

2.8.3.1.4. *Decomposition and Determination Procedure*

2.8.3.1.4.1. *Decomposition.* The weighed sample (30–80 μg) is wrapped in polyethylene film, provided with a piece of linen thread, and wedged into the platinum gauze of the sample holder (cf. Sect. 2.5.3.1.4). 300 μl hypochlorite solution (1 ml measuring pipet) and ca. 300 μl phosphate buffer solution are put into the cleaned combustion flask. The pH value of this solution should lie between 6.0 and 6.5. Therefore the exact amount of phosphate buffer must be determined for each newly prepared sodium hypochlorite solution.[52] The two solutions are thoroughly mixed, the entire wall of the flask except the joint is wet with absorbent solution by rotation of the flask, and the flask is then filled with oxygen. The combustion is carried out as described in Sect. 2.5.3.1.4.

After the combustion the flask is rotated for about 1 min to wet the lower third with absorbent solution. It is then set vertically in an empty desiccator for 30 min, rotated again for 3 min, and left again in the desiccator for 30 min. After the rim of the joint has been wiped with moist filter paper, the stopper is raised a few millimeters and 100 μl distilled water is injected into the gap from a hypodermic syringe. The loosened stopper is rotated several times so that no water is lost. Then the stopper is rinsed with 500 μl water from a hypodermic syringe as the stopper is further withdrawn from the joint, the platinum wire and gauze are carefully rinsed with an additional 500 μl water, and finally the wall is

---

[52] Phosphate buffer solution (x ml) is added to 10 ml hypochlorite solution and 70 ml water until the solution has a pH value of 6.2 (pH meter). The amount of phosphate buffer solution to be introduced for the absorption is then $x - 30$ μl.

rinsed with another 500 μl water. The flask is closed and after 30 min, when the rinse water has drained well off the wall, it is rinsed again with 500 μl water, without wetting a large area of the wall with absorbent solution; after an additional 30 min wait, the determination may begin.

2.8.3.1.4.2. *Determination.* A stirring bar, thoroughly rinsed with water, is put into the solution and the solution is stirred well with the magnetic stirrer of the titration apparatus and then heated 15 min in the aluminum heating block (120 ±1°). Toward the end of the heating period the wall above the solution is rinsed with 100 μl water, about 150 μl sodium formate solution is added, the solution is stirred 30 sec, the flask is set in the heating block again for about 1 min, and it is then, lightly covered, allowed to cool slowly.[53]

After cooling, the wall over the solution is rinsed again with 100 μl water and the solution is stirred well. 15 μl ammonium molybdate solution, ca. 150 μl 12N sulfuric acid, and ca. 150 μl potassium iodide solution are added. After about 1 min the mixture is titrated with 0.01N sodium thiosulfate solution (500-μl piston buret); shortly before the endpoint some Thyodene indicator or starch solution is added, and thiosulfate is added in 0.1-μl increments. The vessel is rotated again to pick up all the iodine.

2.8.3.2. COLORIMETRIC PROCEDURE OF FENNEL AND WEBB FOR MORE THAN 2 μg BROMINE (6)

2.8.3.2.1. *Principle*

Decomposition of the weighed sample (>30 μg) by combustion in the oxygen flask. Colorimetric determination of bromine with cresol red.

2.8.3.2.2. *Equipment*

Kirsten combustion apparatus. Cf. p. 43.
25- and 50-ml volumetric flasks

Spectrophotometer. 20-mm glass cuvets.

Thermostated waterbath

---

[53] It is convenient to insert the narrow, lower part of the flask into a hole in a plastic slab (e.g., PVC, 20 mm thick).

2.8.3.2.3.   *Reagents*

0.001N bromine standard solution. An exactly weighed amount, ideally 119.0_1 mg, of potassium bromide a.r., dried at 110°, is dissolved in some double-distilled water in a 1000-ml volumetric flask; the flask is filled to volume and the titer calculated. The solution is kept in the dark in a polyethylene bottle. 100 μl 0.001N solution corresponds to 7.991 μg bromine.

2N ammonia water. By passing NH_3 from a cylinder into double-distilled water; the solution is kept in a polyethylene bottle.

2N, 9N, and 18N sulfuric acid a.r.

2.5% aqueous sodium nitrite solution. In a brown $\bar{\$}$ glass bottle.

10% aqueous urea solution

Cresol red solution. 30 mg (or 15 or 7.5 mg) cresol red is finely powdered, 0.8 ml 0.1 N sodium hydroxide solution is added, and the solution is shaken until the dye has dissolved completely. The solution is diluted to 1 liter and a part of it is kept in a pipet bottle with a 5-ml pipet.

Potassium bromate solution. 520 mg potassium bromate a.r. is dissolved in 100 ml distilled water; 10 ml of this solution is diluted to 100 ml.

Sodium arsenite solution. 2.5 g As_2O_3 a.r. and 2 g NaOH a.r. are dissolved in 20 ml distilled water; the solution is diluted to 200 ml, neutralized with 1N sulfuric acid (against litmus paper), and diluted to 500 ml. The solution is kept in a glass bottle protected from light.

Oxygen. From a cylinder.

2.8.3.2.4.   *Establishing Calibration Curves in the Ranges of 10–40, 5–20, and 2–10 μg Bromine*

0, 100, 200, 300, 400, and 500 μl 0.001N bromide standard solution are introduced with a 500-μl micro piston buret into a set of 25-ml volumetric flasks; to each is added 500 μl 2N ammonia water, 30 μl sodium nitrite solution, 500 μl 2N sulfuric acid, and 5 ml 9N sulfuric acid. They are stirred for 1 min and 1 ml urea solution is added; they are stirred 4 min more and then

5 ml cresol red solution and finally 1 ml potassium bromate solution are added. The flasks are filled to volume with double-distilled water, closed, shaken well, and put into a water bath at 30 ±0.5° for 60 min.

Then 500 μl sodium arsenite solution is added to each and the flasks are closed and again shaken well. The solutions are transferred to 50-ml volumetric flasks containing 10 ml 18N sulfuric acid, cooled to room temperature, and brought to volume with distilled water.

The absorbances of the bromide-containing solutions are measured against the bromide-free solution in 20-mm cuvets at 519 nm and plotted against the bromide content. A linear calibration curve is obtained.

Calibration curves for 5–20 and 2–10 μg bromine:

The sensitivity of the method is increased by decreasing the dye concentration. 7.5 mg dye per 1000 ml should be used for the 2–10 μg Br range and 15 mg per 1000 ml for the 5–20 μg range. Correspondingly smaller amounts of bromide standard solution are used.

2.8.3.2.5. *Decomposition and Determination Procedure*

2.8.3.2.5.1. *Decomposition.* The sample, containing 10–40 μg (or 5–20 μg or 2–10 μg) bromine, in a little platinum boat, is put into the cavity E of the sample holder of the decomposition apparatus (cf. Fig. 26, p. 44). 500 μl 2N ammonia water is put into the absorbent depression K of the combustion apparatus, which is horizontal in a tube furnace heated to 850°, and the vessel is filled with oxygen by means of a quartz inlet capillary. The sample holder, its joint F moistened with a drop of water, is quickly introduced into the decomposition vessel and the stopper is secured with coil springs. After 1 min the decomposition vessel is removed from the furnace and allowed to cool (ca. 10 min) horizontally. Then the entire wall of the vessel is wet with the absorbent solution, the vessel is left vertical for 5 min and then opened, the sample holder, joint, and wall are rinsed with about 3 ml distilled water, and the rinse water is allowed to drain well (about 20 min).

2.8.3.2.5.2. *Determination.* The contents of the decomposition vessel are quantitatively transferred to a 25-ml volumetric flask (using 3×2 ml rinse water), about 30 μl sodium nitrite solution, 500 μl 2N and 5 ml 9N sulfuric acid are added, and the procedure is continued as described above. The bromine blank value must be determined and subtracted.

2.8.3.3.   PROCEDURE FOR 0.5–5 μg BROMINE (2)

2.8.3.3.1.   *Principle*

The sample (1–10 μg) is weighed by solution partition and pyrolized in a hydrogen stream at about 900°. The hydrogen bromide formed is absorbed in glacial acetic acid and titrated argentometrically. The endpoint is indicated bipotentiometrically.

2.8.3.3.2.   *Equipment*

Decomposition and determination apparatus. The apparatus is similar in all respects to that used for chloride determination (Sect. 2.7.3.2.2). Electrolytically generated hydrogen is needed in place of oxygen (Sect. 2.4.2.3.2).

100–μl displacement buret. Cf. p. 27 ff. For introduction of bromide standard solution.

2.8.3.3.3.   *Reagents*

0.001M bromide standard solution. An exactly weighed amount of potassium bromide "for IR analysis" [54] (about 60 mg) is dissolved in some "halogen-free" triple-distilled water (cf. Sect. 2.7.3.2.3) in a 500-ml quartz volumetric flask [55] which is then filled to volume with water.

The concentration of the solution will not change in the flask for several months if the solution is kept in the dark and protected from dust. 1 μl of an exactly 0.001M solution corresponds to 0.0799 μg bromine. To fill the 100-μl piston buret, the two-necked flask (Fig. 17) described on p. 29 ff, filled daily with standard solution, is used.

---

[54] Before being weighed, the potassium bromide should be dried about 20 hr in a drying oven at 180° and stored over P_2O_5 in a desiccator.

[55] The flask should be cleaned well with concentrated nitric acid and steamed before use for at least 30 min.

Glacial acetic acid. Containing <0.01 μg/ml chloride. Cf. Sect. 2.7.3.2.3.

0.0001N silver nitrate–glacial acetic acid solution. 5.00 ml 0.01N silver nitrate solution (weakly acidified with nitric acid) is diluted to 500 ml with "halogen-free" glacial acetic acid (cf. Sect. 2.7.3.2.3). The solution is stored in a black lacquered polyethylene bottle previously equilibrated with 0.0001N silver nitrate–glacial acetic acid solution.

Acetone. Cf. Sect. 2.7.3.2.3.

Silver bath. Cf. Sect. 2.7.3.2.3.

0.01M potassium bromide solution. For preparation of the silver bromide electrodes.

2.8.3.3.4. *Apparatus Preparation*

An electrolysis time of 1 hr suffices for silvering the cleaned platinum wire electrodes (cf. Sect. 2.7.3.2.4). The silver electrodes are thoroughly rinsed with water while still connected to the current source. They are used for bromide determination in the presence of chloride, iodide, or sulfide (cf. Sect. 2.8.3.3.8). The electrodes are usable for only a few hours.

The silver bromide electrodes needed for bromide determination in the absence of interfering ions (cf. Sect. 2.8.3.3.7) are prepared by connecting the silver electrodes anodically against a platinum wire electrode for 5 min in a 0.01M potassium bromide solution (about 120 μA). They are rinsed free of bromide with double-distilled water while still connected to the current source (practically no current should be flowing) and stored in glacial acetic acid in the dark; the electrodes react at the first titration, but their sensitivity improves during the first several measurements.

A new apparatus is steamed for 1 hr and then rinsed with acetone. The cleaning and filling of the burets is described in Sect. 2.7.3.2.4. After the introduction of the heating coil support and the previously ignited quartz sample support, hydrogen is admitted at about 15 ml/min, the air is displaced from the apparatus, and the coil is brought to about 900°. The pyrolysis vessel is heated until no more halide can be detected in the absorption vessel by titration.

Now several 15-μg portions of a bromine-containing substance are pyrolyzed in order to coat the surface of the apparatus with hydrogen bromide. If all conditions are kept constant, equilibrium

0.10 N sodium thiosulfate solution. For preparation and stand-

If the apparatus is used daily, pyrolysis of one preliminary sample is adequate to reestablish equilibrium.

2.8.3.3.5. *Calibration*

Before and after each daily series of determinations, the silver nitrate solution should be calibrated against bromide standard solution as follows:

With a hydrogen flow of ca. 15 ml/min, 1 ml glacial acetic acid is introduced from the microburet and various amounts (between 5 and 50 μl) of 0.001M KBr standard solution from the 100-μl displacement buret. The electrodes are polarized, the paper transport of the recorder is turned on (2.5 cm/min), and the base potential is allowed to become constant. The titration is begun with a "flying start" (to avoid recorder dead time), i.e., the motor piston buret (about 30 μl/min discharge rate) is turned on at the moment the recorder pen crosses a division line of the paper. The distance between titration start and curve maximum is plotted against the amount of bromide as a calibration curve. This yields a straight line which does not in general pass through the origin (glacial acetic acid blank value).

2.8.3.3.6. *Decomposition Procedure*

The procedure is as in the chlorine determination (cf. Sect. 2.7.3.2.4), except that the apparatus is purged not with oxygen, but with a constant hydrogen flow of 15 ml/min. After the introduction of the sample there is a 4 min wait, then 1 ml glacial acetic acid is admitted into the well-rinsed absorption vessel, stirring is begun, and the Pt–Rh coil is quickly heated to about 900°. During the entire heating time, which must be kept constant, the temperature of the coil is not changed. After 3–4 min the bulk of the hydrogen bromide formed will have gone over into the absorption vessel; a total heating time of 25 min is adequate. It is then necessary to wait about 15 min until the pyrolysis vessel has cooled and the absorbent solution has attained the temperature (e.g., 20°) of the thermostat before beginning the titration.

2.8.3.3.7. *Titrimetric Determination*

In the absence of Cl^-, I^- and S^{2-} with silver bromide electrodes:

After complete absorption of the hydrogen bromide, the electrodes are polarized and the procedure is continued as described in Sect. 2.8.3.3.5, but without introduction of bromide standard solution.

In the presence of Cl^-, I^- or S^{2-}:

Pure silver electrodes are used (cf. Sect. 2.8.3.3.5) which are conditioned before the actual determination by preliminary titrations of about 3 μg bromide (bromide standard solution) until the endpoint for a given amount of bromide is reproducible. Then fresh glacial acetic acid is introduced, the decomposition is performed, the reaction products are transferred, and the titration is carried out as described in Sect. 2.8.3.3.5; the maxima will occur in the order: sulfide, iodide, bromide, chloride. In this case the titration rate should be between 10 and 20 μl/min. A subsequent titration is performed, for which fresh glacial acetic acid is introduced and a known amount of bromide standard solution, corresponding roughly to the amount of bromide found, is added, and the two determinations are correlated.

2.8.4. REFERENCES

1. H. ter Meulen and A. Slooff, *Rec. Trav. Chim.*, **59**, 259 (1940).
2. W. H. List and G. Tölg, *Z. Anal. Chem.*, **226**, 127 (1967).
3. R. Belcher, Y. A. Gawargious, P. Gouverneur, and A. M. G. Macdonald, *J. Chem. Soc.*, **1964**, 3560.
4. R. Belcher, R. A. Shah, and T. S. West, *J. Chem. Soc.*, **1958**, 2998.
5. I. M. Kolthoff and H. Yutzy, *Ind. Eng. Chem., Anal. Ed.*, **9**, 75 (1937).
6. T. R. F. W. Fennell and J. R. Webb, *Z. Anal. Chem.*, **205**, 90 (1964).
7. M. J. Fishman and M. W. Skougstad, *Anal. Chem.*, **35**, 146 (1963).
8. J. D. Winefordner and M. Tin, *Anal. Chem.*, **35**, 382 (1963).
9. J. T. Stock, *Amperometric Titrations*, Interscience, New York, 1965.

10. E. Bishop and R. G. Dhaneshwar, *Analyst,* **87**, 845 (1962).
11. I. Čadersky, *Mikrochim. Acta,* **1966**, 401.
12. G. Schwab and G. Tölg, *Z. Anal. Chem.,* **205**, 29 (1964).
13. R. P. Larsen and N. M. Ingber, *Anal. Chem.,* **31**, 1085 (1959).

2.9.   IODINE

2.9.1.   DECOMPOSITION, GENERAL

For iodine determination, combustion of the sample in pure oxygen (1–5) and hydrogenating decomposition (6–8) yield sufficiently small blank values. Decomposition in the oxygen flask, modified for samples of 30–100 μg (9) and 1–30 μg (10), is especially simple. Polyethylene film is used as combustion support in the first case and paper is used in the second. Decomposition with metallic potassium (11) is less advisable.

2.9.2.   DETERMINATION, GENERAL

Of the numerous methods for determination of small amounts of iodine (10, 12–19) the following are especially accurate.

2.9.2.1.   THIOSULFATE TITRATION (STARCH INDICATOR)

According to Leipert (20), amounts of iodine over 10 μg/3 ml volume can be determined with 0.01 N thiosulfate solution and starch indicator with a standard deviation of about ± 0.1 μg iodine. The method is used in improved form (11) by Belcher, Gawargious, Gouverneur, and Macdonald (9) after combustion of the sample in the oxygen flask. The iodine is oxidized to iodate by bromine water in neutral solution and the excess bromine is reduced by addition of formic acid.

The method is not suitable for significantly smaller amounts of iodine.

2.9.2.2.   IODOMETRIC TITRATION WITH BIAMPEROMETRIC ENDPOINT INDICATION (10)

The procedure of Potter and White (21) was applied to the ultramicro scale.

After the decomposition, iodide and iodine are converted to iodate. Excess iodide is then added, the liberated iodine is reacted with a known amount of thiosulfate, and the unconsumed thiosulfate is back-titrated with iodate solution. The circuit for biamperometric endpoint indication is shown in Fig. 57. The platinum elec-

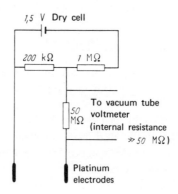

Figure 57. Circuit for biamperometric titration.

trodes are polarized with 300 mV. The electrode current is measured with a vacuum tube voltmeter by the potential drop across a 50-Mohm resistance. Since the relatively large diffusion current of oxygen would greatly reduce the sensitivity of the indication, the solution must be purged with oxygen-free nitrogen for about 5 min before the titration. The nitrogen bubbling through the solution also serves to mix the solution during the titration. For titration of 0.5 μg I, the standard deviation of the method is about ±2 ng.

2.9.2.3. ARGENTOMETRIC DETERMINATION WITH DITHIZONE INDICATOR (10)

In the titration of iodide with Ag^+ ions in sulfuric acid solution in the presence of dithizone, only the excess silver ions react with dithizone forming very intensely yellow primary silver dithizonate (AgHDz), causing the color to change from green to yellow (22). Silver chloride and silver bromide are less stable than AgHDz, so chloride and bromide do not interfere with the determination. Since dithizone and silver dithizonate are practically insoluble in water, titration can be performed only in the presence of an organic solvent, e.g., acetone (one phase) or $CHCl_3$, CCl_4, or other solvent insoluble in water (two phases). Two-phase titration produces an appreciable increase in sensitivity of endpoint indication over one-phase titration. It is possible to operate in such a way that the volume of the organic phase (indicator phase) amounts to only a fraction of the aqueous phase. First (excess iodine) the indicator phase is green; with an excess of silver it turns yellow immediately if the two phases are mixed well during the titration. For the color change to be visible during the mixing, a part of the heavy organic phase must remain on the floor of the titration vessel. Since the amount of dithizone used as indicator consumes an equivalent amount of silver ions before changing color, the sensitivity will be greatest if the amount of dithizone is kept so small that it is just possible to recognize a change from green to yellow.

Of the other metals, only Pt^{2+}, Pd^{2+}, Au^{3+}, and Hg^{2+}, which react with dithizone before Ag^+ (22), interfere in sulfuric acid solution. These ions, however, can hardly be expected as interfering impurities in the reagents; only platinum can be introduced during the decomposition. Of the anions in question, only sulfide, which reacts stoichiometrically with Ag^+, interferes. Bromide, chloride, and cyanide interfere only when in considerable excess. In the determination of 1 μg iodide, a standard deviation of $s_{25} = \pm 5$ ng was obtained.

2.9.2.4. SPECTROPHOTOMETRIC DETERMINATION WITH CRYSTAL VIOLET (10)

While the simple iodides of the basic triphenylmethane dyes can be extracted only poorly from neutral aqueous solution into benzene or toluene, salts with the polyanions I_2Cl^- or ICl^- go into the organic phase from acid solutions almost completely (23,24). To form the complex anion, iodine must be present as the element; this is obtained by oxidation of iodide by, e.g., nitrite. According to Hillmann and Kuhlmann (23), in the most favorable system, crystal violet – benzene, 1 μg I produces an absorbance of about 0.2 if the compound is extracted with 5 ml benzene (10-mm cuvets).

The present author's investigations (10) showed that the method, applied to smaller amounts of iodine, yielded a relative standard deviation of $\pm 3.5\%$ for 0.64 μg I. Changing the iodide oxidation made possible the design of a more sensitive and accurate procedure.

Iodide is not oxidized to iodine by nitrite, but to iodate by bromine water; treated with iodide, the iodate produces an amount of iodine six times the original. The added iodide interferes only by its impurities of iodate and free iodine, which are quite slight, however, for freshly prepared reagent stored in the dark. For the determination of 1 μg I$^-$ in 2 ml aqueous phase, ca. 100 μl 0.01 M KI solution is necessary.

To reduce the partition of the pure dye into the benzene, the crystal violet concentration of the aqueous phase is reduced to 0.0025 M. Because of the changing distribution of the pure dye, a change in dye concentration in the 0.001–0.01 M range produces no change in absorbance for a given amount of iodide.

Instead of hydrochloric acid (23), it is better to use sulfuric acid, which reacts more slowly with the dye. The most favorable acid concentration is in the range of 0.25–0.6 N (cf. Fig. 58): the blank value is not troublesome and the complex is stable for up to 15 min of extraction. After phase separation by centrifugation, the color of the extracted salt in the benzene phase is constant for about 15 min in indirect sunlight and for several hours in the dark.

Since chloride ions are necessary for the formation of the complex iodide anion, the dye should be added as hydrochloride or the appropriate concentration of chloride ions added.

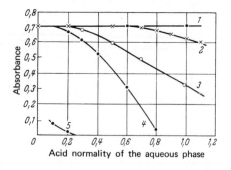

Figure 58. Effect of acid concentration of the aqueous phase on absorbance of the benzene phase in the crystal violet procedure. Conditions: 0.72 μg I$^-$, 2 ml aqueous phase, 2 ml benzene, 20 μl 0.01 M potassium iodide solution, 2 min shaking time, measured at 610 nm against iodine-free reference solution, 10 mm cuvets. Curve 1: 0.005 M, crystal violet, sulfuric acid; curve 2: 0.0025 M, crystal violet, sulfuric acid; curve 3: 0.0005 M, crystal violet, sulfuric acid; curve 4: 0.0005 M, crystal violet, hydrochloric acid; curve 5: blank for 2.

After the oxidation of iodide to iodate with bromine, the removal of traces of bromine by phenol is possible if the amount of phenol is kept as small as possible. Crystal violet phenolate, when it is present in substantial amounts, likewise goes into the benzene phase better as the acid concentration is decreased. Therefore the specifications of amounts and concentrations of reagents should be heeded accurately. The calibration curve (cf. Fig. 59) obtained from the given procedure (cf. Sect. 2.9.3.4.4) is linear in the range of 0.1–1.2 μg I (2 ml benzene and 10-mm cuvets). The procedure yielded a standard deviation of $s_{20} = \pm 8$ ng I for 0.72 μg I.

2.9.3. PROCEDURES FOR IODINE DETERMINATION

2.9.3.1. PROCEDURE OF BELCHER, GAWARGIOUS, GOUVERNEUR, AND MACDONALD (9) FOR MORE THAN 5 μg IODINE

2.9.3.1.1. *Principle*

After decomposition of the weighed sample (30–80 μg) in the oxygen flask, the iodine is oxidized to iodate by bromine. The iodine liberated by the reaction of iodate with iodide is titrated with thiosulfate (starch indicator).

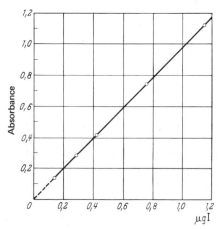

Figure 59. Calibration curve for spectrophotometric crystal violet procedure. Conditions: iodide–iodate oxidation, 2 ml 0.5 N sulfuric acid, 0.0025 M crystal violet solution, 50 μl 0.01 M KI solution, 2 ml benzene, 2 min extraction, 610 nm, 10 mm cuvets, 25°.

2.9.3.1.2. *Equipment*

Combustion flask. Cf. p. 42 ff.

Polyethylene film and linen thread. Cf. p. 43.

Titration apparatus. Cf. p. 24 ff.

500-μl micro piston buret. Cf. p. 24 ff.

2.9.3.1.3. *Reagents*

0.01 N sodium thiosulfate solution. For preparation and standardization, cf. Sect. 2.9.3.1.4.2.

Saturated bromine water. In a glass pipet bottle.

90% formic acid a.r. In a glass pipet bottle.

0.05% methyl red solution (sodium salt). In a glass pipet bottle.

Approximately 2 N sodium hydroxide solution. In a glass pipet bottle.

Approximately 2 N sulfuric acid

10% potassium iodide solution. The aqueous solution must be prepared fresh daily; it is stored in a brown glass bottle with a pipet.

Starch solution or Thyodene (Purkis, Williams Ltd., London)

Single-distilled water from a glass still is adequate for the preparation of all the aqueous solutions.

2.9.3.1.4. *Decomposition and Determination Procedure*

2.9.3.1.4.1. *Decomposition.* The weighed sample (30–80 μg) is wrapped in polyethylene film (cf. Sect. 2.5.3.1.4) with a piece of linen thread and wedged into the platinum gauze of the sample holder. About 100 μl water is introduced into the combustion flask and distributed over the wall by rotation. Subsequently 100 μl 2 *N* sodium hydroxide solution is added to the liquid on the floor of the flask. Then the flask is filled with oxygen and the combustion is carried out (cf. Sect. 2.5.3.1.4). Afterward the closed flask is rotated for 1 min to wet the lower third of the spherical wide section with absorbent solution. The flask is set aside in a vertical position for 30 min, again rotated for 3 min, and again left in dust-free air for 30 min. The joint rim is then wiped with moist filter paper, the sample holder is removed, and, as described in Sect. 2.8.3.1.4.1, rinsed first with 1.5 ml water and after 30 min drainage with 0.5 ml water, without wetting much of the wall with the alkaline absorbent solution. The rinse water is allowed to drain off the wall an additional 30 min.

2.9.3.1.4.2. *Determination.* After the last waiting period, a small, thoroughly rinsed stirring bar is put into the solution with forceps, 20 μl methyl red solution is added and the solution is mixed well and neutralized with 2 N sulfuric acid from a 500-μl micro piston buret. As the buret tip is removed from the solution, it is rinsed with 100 μl water. Then about 150 μl bromine water is added, the solution is stirred 5 min, about 120 μl formic acid is added, and stirring is continued 10 min, during the last few minutes of which the remaining bromine vapors are carefully sucked out of the flask.

After addition of about 50 μl 2 N sulfuric acid and 150 μl potassium iodide solution, the solution is titrated with 0.01 N sodium thiosulfate solution; some starch indicator is added toward the end of the titration, and the titration is finished with 0.1-μl additions. In a blank determination, not more than 2–3 μl thiosulfate solution should be consumed, corresponding to about 0.5 μg I.

2.9.3.2. IODOMETRIC PROCEDURE FOR LESS THAN 2 μg IODINE (10)

2.9.3.2.1. *Principle*

The sample is measured out by solution partition and burned in the oxygen flask on a paper support. After conversion of the iodine to iodate it is reacted with iodide and the liberated iodine is titrated with thiosulfate solution. The endpoint indication is biamperometric.

2.9.3.2.2. *Equipment*

Three 500-μl piston burets with reservoirs. Cf. p. 24 ff.

Combustion flask. Cf. Fig. 27 and p. 43 ff.

Teflon stopper with platinum wire electrodes.   Cf. p. 44.

Induction coil: About 10 kV (e.g., automotive ignition coil with interrupter).

Heating block with thermometer. Cf. Sect. 1.3.3.

Titration apparatus. Cf. Figs. 14 and 60. A ⚶ 14.5 polyethylene or Teflon stopper, fastened to a stand by a holder, has five holes (2 mm diameter). Polyethylene tubes, both drawn out to capillaries, pass through two holes. One tube is for admission of nitrogen and the other is connected to a micro piston buret with reservoir. Two platinum wire electrodes (0.3 mm diameter, 3 mm effective length), fused in glass tubing (2 mm diameter), pass through two other holes. The fifth hole is for nitrogen exit.

For titration, the apparatus is inserted into the combustion flask. The vacuum tube voltmeter [56] (10- and 100-mV ranges) for

---

[56] For example, vacuum tube voltmeter with amplifier and compensation of Frieseke and Hoepfner Co., Erlangen- Bruck.

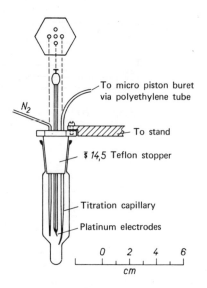

Figure 60. Titration apparatus for iodine determination with biamperometric endpoint indication.

the biamperometric circuit (cf. Fig. 57) must have an internal resistance > 100 Mohm and zero suppression.

Before use, the platinum electrodes are etched briefly with aqua regia, ignited, and prepared by electrolysis (6 V) in 2 N sulfuric acid, connected alternately as cathodes and anodes against an auxiliary electrode for several minutes.

2.9.3.2.3. *Reagents*

Freshly double-distilled water from a quartz still is used for the preparation of all solutions.

Iodate standard solutions. 0.1 N stock solution: 3.567 g potassium iodate a.r. (dried 2 hr at 120°) is dissolved in water and diluted to 1000 ml. From this, 0.001 N and 0.0001 N working solutions are prepared daily (cf. Sect. 1.2.6).

Thiosulfate solutions. 0.1 N stock solution: stored in the dark. From it, 0.001 and 0.0001 N solutions are prepared daily (cf. Sect. 1.2.6).

Iodide standard solutions. Cf. Sect. 2.9.3.3.3.

Approximately 6% acetic acid. 6 ml glacial acetic acid a.r. is diluted with water to 100 ml.

Potassium iodide a.r.

Bromine water. Double-distilled water saturated with bromine a.r., kept in a ⚗ bottle with a 20-μl pipet.

0.5% aqueous phenol solution. Stored in a brown ⚗ glass bottle.

Nitrogen. Postpurified cylinder nitrogen.

2.9.3.2.4. *Decomposition and Determination Procedure*

2.9.3.2.4.1. *Decomposition.* The dissolved sample is transferred to a paper disk of 5 mm diameter (as described on pp. 44 ff) which is clamped in the platinum loop of the electrode holder shown in Fig. 27, *b*. 500 μl double-distilled water and 20 μl saturated bromine water are introduced into the combustion flask and it is filled with purified oxygen (a powerful oxygen stream of about

30 sec duration). It is then closed with the Teflon stopper of the electrode holder. The wall of the combustion flask is wet by rotating it. All operations are carried out in a glove box (cf. Fig. 3).

The paper disk is ignited by an induction coil. To absorb the combustion products, the flask is shaken for 10 min by a shaking machine or left standing for at least 30 min (10) with intermittent wetting of the wall with absorbent solution (cf. Fig. 61).

For difficultly combustible substances, such as sulfonic acids and their salts, paper disks prepared with picric acid produce higher combustion temperatures and hence more complete combustion of the sample (10).

2.9.3.2.4.2. *Determination.*

The combustion flask is opened for the titrimetric determination. The Teflon stopper is first raised only 1–2 mm and the stopper and joint are meticulously rinsed with about 1 ml double-distilled water with an injection syringe or a rubber bulb pipet. Then the electrode leads and platinum wires are rinsed with an additional milliliter of double-distilled water and the stopper is laid aside. Now the excess bromine must be completely removed from the solution. This can be done by evaporating the solution (about 2.5 ml) by at least half while bubbling a fine stream of nitrogen from a glass capillary through it (cf. Fig. 29). To avoid iodine losses, the decomposition flask should be heated to not

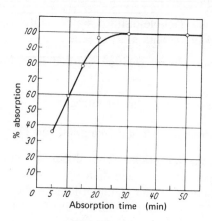

Figure 61. Time dependence of iodine absorption after combustion: 500 μl 1 *N* NaOH absorbent solution; 1 μg iodine.

over 110° at only its lower third (metal heating block) so that no residue can form on the wall.

A second method for removal of bromine consists of subsequent addition of phenol (about 20 μl of 0.5% solution). The solution containing the bromine is heated for only about 15 min in the heating block, during which nitrogen bubbling is not necessary, and then the phenol solution is added. Afterward the wall of the combustion flask (except the joint) is wet with the solution and the solution is left to cool. The trace of bromine still present corresponds to less than 2 ng iodine. Iodine losses greater than 2 ng were not observed. The amount of phenol may be raised by a factor of five.

After removal of the bromine, 1 ml of ca. 6% acetic acid is added to the absorbent solution (about 1 ml), carefully rinsing the vessel wall with the acetic acid, and then 100 ± 0.1 μl 0.0001 N thiosulfate solution is introduced. Then about 15 mg solid potassium iodide is added and the titration apparatus is inserted into the decomposition flask. After the solution has been purged for 5 min with nitrogen, the titration is begun while the nitrogen bubbling continues. 0.0001 N potassium iodate solution is added from a micro piston buret until the needle of the measuring instrument deflects. Then smaller increments are added, with the current or potential change being measured 30 sec after each addition. A plot of potential versus potassium iodate consumption yields two straight lines, the intersection of which indicates the titration endpoint (cf. Fig. 62).

If the scatter of the measured points is too great, a computational evaluation is advisable (25).

Since the titer of a 0.0001 N thiosulfate solution is constant for only about 12 hr, it is best to standardize before and after each determination. Furthermore, the blank value of the reagents must be taken into consideration. The procedure can be simplified by using a motor-driven buret and a millivolt recorder.

2.9.3.3. ARGENTOMETRIC PROCEDURE FOR LESS THAN 2 μg IODINE (10)

2.9.3.3.1. *Principle*

The sample is measured out by solution partition onto a paper

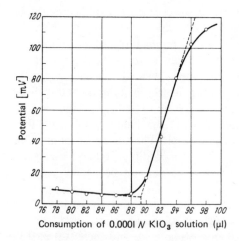

Figure 62. Curve for the biamperometric titration of 100 μl 0.0001 N thiosulfate solution with 0.0001 N KIO$_3$ solution: titration endpoint: determined graphically: 89.4μl; calculated: 89.3 μl (25).

support and burned in the oxygen flask. After addition of ascorbic acid the iodide is titrated argentometrically. Dithizone serves as indicator.

2.9.3.3.2. *Equipment*

Two 500-μl micro piston burets with reservoirs (cf. p. 24 ff)

Combustion flask. Cf. p. 43 ff and Fig. 27, *a*.

Teflon stopper with electrodes. Cf. p. 44.

Induction coil. 10 kV (e.g., automotive ignition coil with interrupter.

Heating block with thermometer. Cf. Sect. 1.3.3.

Titration apparatus. Cf. Figs. 14 and 63 and p. 24. A ⍑ 14.5 polyethylene or Teflon stopper, fastened to a stand by a holder, has three holes of 2 mm diameter. A spiral stirrer (other types have not proved satisfactory) passes through the center hole and is driven at variable speed by a small, high speed dc motor.
Through the second hole passes a polyethylene tube (2 mm

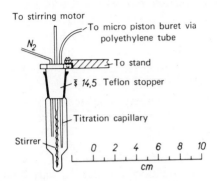

Figure 63. Apparatus for two-phase titration.

diameter), the lower end of which is drawn out to a capillary and the upper end connected to a micro piston buret. Through the third hole passes another polyethylene tube drawn out to a capillary, which is for admission of nitrogen. The titration vessel is attached to the stopper from below; its shape permits rapid phase separation and good recognition of color change in the narrow section (Fig. 63).

2.9.3.3.3. *Reagents*

Iodide standard solution. Stock solution: 166.0_1 mg [57] potassium iodide a.r. (dried in the dark for 24 hr over phosphorus pentoxide in a vacuum desiccator) is dissolved in some water and diluted to 1000 ml; the solution is kept in a black lacquered quartz bottle. The working solution is prepared fresh daily by dilution of the stock solution in a 1 : 10 ratio. 1 μl corresponds to 12.6_9 ng iodide.

Silver standard solution. Stock solution: 169.6_7 mg [57] silver nitrate a.r. (dried 24 hr over phosphorus pentoxide in a vacuum desiccator) is dissolved in 10 ml very pure 2 N sulfuric acid and some water; the solution is diluted with water to 1000 ml and stored in a black lacquered polyethylene bottle. 0.0001 N working solution: prepared fresh daily by 1 : 10 dilution of the stock solution. 1 μl corresponds to 12.6_9 ng iodide.

---

[57] Since it would be very time consuming to weigh this amount exactly, it is weighed approximately and a titer is used.

2 N sulfuric acid solution. Water is added to 28 ml concentrated sulfuric acid, redistilled with a quartz still, and this is diluted to 500 ml. The acid is kept in a quartz bottle.

0.05% dithizone stock solution. 50.0 mg dithizone a.r. is dissolved in 100 ml carbon tetrachloride ("suitable for analysis with dithizone"). The solution, stored in the dark, is usable for about 3 weeks. Working solution: The stock solution is diluted 1 : 100 with CCl_4. The solution, stored in the dark, is usable for a maximum of 8 hr.

5% aqueous ascorbic acid solution. 5 g ascorbic acid a.r. and 95 ml water. The solution will keep for only 2 days.

Ca. 1 N sodium hydroxide solution. 4 g NaOH a.r. is dissolved in water in a 100-ml quartz volumetric flask and diluted to volume. The solution is kept in a polyethylene pipet bottle with a 500-μl pipet.

All aqueous solutions are prepared with water double-distilled over silver powder in a quartz still.

Purified nitrogen. Cf. Sect. 2.3.2.1.2.

2.9.3.3.4. *Silver Nitrate Solution Standardization*

100.0 μl 0.0001 N iodide standard solution (1.269 μg I) is introduced from a 500-μl piston buret and 1 ml double-distilled water, 1 ml 2 N sulfuric acid, 500 μl 5% ascorbic acid solution, and 50.0 μl dithizone solution [58] are added. The flask is attached to the Teflon stopper of the titration apparatus, with the capillary for nitrogen admission not dipping into the solution. The two phases are vigorously mixed by means of the stirring motor, but a small drop of the carbon tetrachloride phase should always remain on the floor of the decomposition flask. Silver nitrate is slowly added from the second micro piston buret until the green solution begins to turn yellow. To recognize the color change better, a

---

[58] With 3 ml aqueous phase, the volume of the CCl_4 phase should not be smaller than 50 μl, for otherwise too large a fraction of the CCl_4 would remain distributed in the aqueous layer when the phases are mixed. The 20 $\mu mole$ of dithizone consumes silver solution corresponding to 129 ng iodide. If the dithizone concentration were decreased, the color change would be hard to recognize.

sheet of white paper is held behind the titration vessel and the stirring is momentarily stopped if necessary. The silver nitrate blank consumption is similarly determined and subtracted before the silver nitrate solution titer is calculated.

If the titration is carried out properly, the silver nitrate solution titer can be determined with a relative standard deviation of <0.4%.

2.9.3.3.5. *Decomposition and Determination Procedure*

2.9.3.3.5.1. *Decomposition.* The decomposition is carried out according to the procedure of Sect. 2.9.3.2.4.1, except that 200 μl 1 N sodium hydroxide solution is employed in place of water and bromine water, without wetting the wall.

2.9.3.3.5.2. *Determination.* When the combustion products have been absorbed, the joint, stopper, electrodes, and wall are scrupulously rinsed with 1 ml double-distilled water (injection syringe); then 1 ml 2 N sulfuric acid is added, which is also used to rinse the wall. The combustion vessel is attached to the stopper of the titration apparatus and nitrogen is bubbled through the solution for 10 min. Otherwise the oxygen dissolved in the absorbent solution and the ozone originating in the spark ignition would immediately bleach the dithizone solution. Postpurified nitrogen from a steel cylinder is usually pure enough, but the nitrogen must be admitted into the solution only via a polyethylene tube well rinsed with water, never via a rubber tube. The nitrogen inlet capillary is withdrawn from the solution and the procedure is continued as described in Sect. 2.9.3.3.4.

The blank value is determined analogously and taken into account.

2.9.3.4. SPECTROPHOTOMETRIC PROCEDURE FOR LESS THAN 2 μg IODINE (10)

2.9.3.4.1. *Principle*

After the sample has been measured out by solution partition, it is decomposed in the oxygen flask and all iodine is converted to iodate. The iodine liberated upon addition of potassium iodide is extracted with benzene as crystal violet iodochloride and the absorbance of the benzene phase is determined spectrophotometrically.

2.9.3.4.2. *Equipment*

Combustion flasks. Ca. 10-ml capacity with ⟟ 14.5 joint. In the shape of centrifuge tubes. Fitting polyethylene stoppers.

Teflon stopper with electrodes. Cf. p. 44 and Fig. 27, *b*.

Induction coil. 10 kV, e.g., automotive ignition coil with interrupter.

Heating block with thermometer and several holes. Only the lower third of the combustion flasks can be inserted into the holes (cf. Fig. 29).

Spectrophotometer. With two 1-cm or 2-cm glass cuvets with tight Teflon stopper closures.

Device for transfer of organic solvents. Cf. p. 22, Fig. 10.

Thermostat

Centrifuge

2.9.3.4.3. *Reagents*

Iodide standard solution. Cf. Sect. 2.9.3.3.3.

2N Sulfuric acid. Cf. Sect. 2.9.3.3.3. In 2-ml microburet with reservoir.

0.005 M crystal violet solution. 2.14 g crystal violet a.r. dissolved in 1000 ml double-distilled water; if the crystal violet is not present as the hydrochloride, a corresponding amount of NaCl is added. The solution is measured out with a 1-ml microburet with reservoir.

Ca. 0.01 M potassium iodide solution. 16.6 mg KI a.r., dissolved in 10 ml double-distilled water; the solution is prepared fresh daily and is kept in the dark in a pipet bottle with a 50-μl pipet.

0.1% phenol solution. Phenol a.r., dissolved in double-distilled water; the solution is kept in a brown ⟟ bottle with a 100-μl pipet inserted.

Bromine water. Cf. Sect. 2.9.3.2.3.

Benzene a.r. Kept in a 2-ml microburet with Teflon plug.

2.9.3.4.4. *Establishing the Calibration Curve*

Accurately measured amounts of the iodide standard solution (between 10 and 100 μl, corresponding to 0.126–1.26 μg I $^-$) are introduced from a micro piston buret into six centrifuge tubes with $\overline{\mathbb{S}}$ 14.5 joints, previously cleaned thoroughly with hot concentrated nitric acid and double-distilled water and then steamed 10 min. 500 μl double-distilled water and 20–30 μl saturated bromine water are added to each. After 10 min the excess bromine is removed by heating the lower third of the centrifuge tubes for 15 min in an aluminum heating block at not over 110°. The solution is allowed to cool, 100 μl 0.1% phenol solution (never more) is added, and the solution is mixed. Then the following reagents are added in the order: 500 μl 2 N sulfuric acid, 50.0 μl 0.01 M potassium iodide solution, 2.00 $\pm$ 0.002 ml benzene (by means of the pipetting apparatus), and 1 ml 0.005 M crystal violet solution (semimicroburet).

The vessels are tightly closed with polyethylene stoppers and each tube is immediately shaken for 2 min by hand. Then the solutions are cooled 2–3° below room temperature and centrifuged for 2 min at > 2000 rpm. The organic phases are quickly transferred, one after another, into the cuvets by means of the pipetting apparatus; the cuvets are immediately closed tightly and the absorbances of the solutions are measured within an hour against pure benzene at 610 nm. Because of the large coefficient of thermal expansion of benzene, thermostating of the cuvet chamber of the spectrophotometer and of the benzene buret is necessary for accurate measurements. After each use, the cuvets should be thoroughly cleansed of dye remains, which adhere very tightly to glass, with concentrated nitric acid. The same applies to the decomposition tubes and the pipetting apparatus.

2.9.3.4.5. *Decomposition and Determination Procedure*

2.9.3.4.5.1. *Decomposition.* For combustion in the oxygen flask (cf. Sect. 2.9.3.2.4.1), 100 μl double-distilled water and 20–30 μl bromine water are introduced. The platinum wires of the ignition device must never come in contact with the bromine solution, for dissolved traces of platinum could interfere with the determination. Furthermore, the ignition sparks should arc over as briefly as possible. The electrodes and joints are rinsed after the combustion and absorption with 0.5 ml water (rubber bulb pipet with fine tip or injection syringe).

2.9.3.4.5.2. *Determination.* The photometric determination is carried out as described in Sect. 2.9.3.4.4.

2.9.4. REFERENCES

1. R. Belcher and J. E. Fildes, *Analyst,* **10**, 383 (1961).
2. K. Hozumi and K. Mizuno, *Japan Analyst,* **10**, 383 (1961).
3. E. Pella, *Mikrochim. Acta,* **1962**, 916.
4. W. Schöniger, *Z. Anal. Chem.,* **181**, 28 (1961).
5. A. M. G. Macdonald, *Analyst,* **86**, 3 (1961).
6. A. Slooff, *Rec. Trav. Chim.,* **59**, 259 (1940).
7. W. J. Kirsten, *Z. Anal. Chem.,* **181**, 1 (1961).
8. M. N. Chumachenko and V. P. Miroshina, *Zavod. Lab.,* **26**, 1084 (1960).
9. R. Belcher, Y. A. Gawargious, P. Gouverneur, and A. M. G. Macdonald, *J. Chem. Soc.,* **1964**, 3560.
10. B. Morsches, and G. Tölg, *Z. Anal. Chem.,* **200**, 20 (1964).
11. R. Belcher, R. A. Shah, and T. S. West, *J. Chem Soc.,* **1958**, 2998.
12. V. Stolč, *Mikrochim. Acta,* **1961**, 710; **1963**, 941.
13. H. V. Malmstadt and T. P. Hadjiioannou, *Anal. Chem.,* **35**, 2157 (1963).
14. E. Jungreis and J. Gedalia, *Mikrochim. Acta,* **1960**, 145.
15. E. I. Savicher, I. G. Vasil'era, and E. I. Golovin, *Zavodsk. Lab.,* **29**, 1433 (1963).
16. K. B. Yatsimirskiǐ, L. I. Budarin, N. A. Blagoveshchenskaya, R. V. Smirnova, A. P. Fedorova, and V. K. Yatsimirskiǐ, *Zh. Anal. Khim.,* **18**, 103 (1963).
17. S. Utsumi, M. Shiota, N. Yonchara, and I. Iwasaki, *J. Chem. Soc. Japan, Pure Chem. Sect.,* **85**, 32 (1964).
18. Y. Yamamoto and S. Kinuwaki, *Bull. Chem. Soc. Japan,* **37**, 434, (1964).
19. E. Ramanauskas, *Khim. Tekhnol.,* **1964**, 9.
20. T. Leipert, *Biochem. Z.,* **261**, 436 (1933); *Mikrochim. Acta,* **1938**, 73, 147.

21. E. C. Potter and J. F. White, *J. Appl. Chem.,* **7**, 309 (1957).
22. G. Iwantscheff, *Das Dithizon und seine Anwendung in der Mikro-
    und Spurenanalyse,* Verlag Chemie, Weinheim/Bergstrasse, 1958.
23. G. Hillmann and E. Kuhlmann, *Z. Physiol. Chem.,* **331**, 109
    (1963).
24. L. M. Lapin and N. V. Rejs, *Lab. Delo,* **7**, No. 8, 21 (1961); cf.
    Z. Anal. Chem., **189**, 432 (1962).
25. J. Mika, *Die Methoden der Mikromassanalyse, Die Chem.
    Analyse,* Vol. 42, Ferd. Enke-Verlag, Stuttgart, 1958.

2.10. PHOSPHORUS

2.10.1. DECOMPOSITION, GENERAL

The usual combustion technique in the oxygen flask (1), in which the solid sample is wrapped in a substantial amount of paper, is suitable for the determination only of more than about 10 μg phosphorus, for the phosphorus blank value of paper cannot be reduced appreciably below 0.05 μg/cm^2 even after purification. Therefore polyethylene film, which has no perceptible blank value (2), is suggested for wrapping the sample.

Less than 5 μg phosphorus can be determined, after solution partition of the sample, with a standard deviation of about ± 4 ng P (3) if the area of the paper support used is not greater than 0.2 cm^2 and is kept constant (cf. p. 44). The method yields somewhat low values and is relatively complicated. Decomposition with perchloric acid is more advantageous (1,3–5).[59] It is immaterial whether this is carried out in the presence of concentrated sulfuric acid or concentrated nitric acid, as long as the decomposition temperature is controlled and the mixture is never evaporated to dryness. For less than 5 μg phosphorus, quartz vessels are necessary. The decomposition must be conducted in the absence of dust with specially purified acids. Phosphate-containing rinsing or cleaning agents must never be used in the laboratory.

2.10.2. DETERMINATION, GENERAL

For the determination of very small amounts of phosphorus, photometric molybdenum blue procedures (2,3) ($\epsilon = \sim 23,000$) are most suitable, while somewhat larger amounts can also be titrated (2) by precipitating and acidimetrically determining quinoline phosphomolybdate (6,7). One-step molybdenum blue procedures (8,9), in which low-valence molybdenum of known reducing value is already present in the reagent, are less susceptible to interference (10,11) than two-step procedures, in which the phosphomolybdic acid first formed is reduced in a second step. However, prior extraction of the phosphomolybdic acid with a

---

[59] A number of phosphoric acid esters cannot be broken down by evaporation with oxidizing acids, thus necessitating combustion in oxygen.

water-insoluble organic solvent (1) before the reduction also yields high accuracy.

A substantial increase in sensitivity ($\epsilon = 150,600$) is possible (12) by indirect photometric determination of phosphorus through molybdenum obtained from the extracted molybdophosphoric acid (e.g., as molybdenum(V) thiocyanate).

For 15 determinations of 1 μg phosphorus the one-step Zinzadze procedure (8) yielded a standard deviation of $\pm 0.6\%$ P; for the same number of determinations with a two-step extraction procedure (1), the relative standard deviation amounted to $\pm 1.1\%$ P. A volume of 1 ml and an optical path of 1 cm were used in both cases.

The inconvenient reagent preparation of the original Zinzadze procedure (8) can be replaced by a very simple procedure (9), but the reagent is then less stable.

The molybdenum blue complex can be extracted completely with several organic solvents, e.g., methyl isobutyl ketone, which displaces the absorption maximum from 825 to 780 nm. The complex in the organic phase is quickly bleached in direct sunlight, but hardly at all in the dark (cf. Fig. 64), assuming very pure methyl isobutyl ketone.

Thus extraction permits a subsequent decrease of volume with only a slight loss of accuracy. For the determination of 0.5 μg phosphorus (1-ml determination volume, 10-mm optical path), the standard deviation of the method is $s_{10} = \pm 5$ ng = 1%. Since the

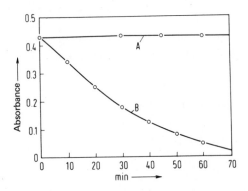

Figure 64. Time dependence of the absorbance of molybdenum blue complex in methyl isobutyl ketone: *A*, in the dark; *B*, indirect sunlight.

extraction method obviates establishing the exact volume of the aqueous phase (no volumetric flask necessary), this technique is superior to photometry of the aqueous solution for very small volumes. For the determination of 1–10 μg phosphorus, Belcher, Macdonald, Phang, and West (2) recommend a procedure of Chalmers and Thomson (13) which uses ammonium iron(II) sulfate as reducing agent.

More than 5 μg phosphorus can be determined very accurately (4) via the colored molybdovanado complex (430 nm absorption maximum). This procedure is about 20 times less sensitive than the molybdenum blue procedure, but simpler to perform. With absorption measurements in the UV range ($\lambda = 310$ nm) a molar absorptivity ($\epsilon = 22400$) is attained which is practically equal to that of the molybdenum blue procedures.

2.10.3. PROCEDURES FOR PHOSPHORUS DETERMINATION

2.10.3.1. PROCEDURE OF SALVAGE AND DIXON (4) FOR MORE THAN 5 μg PHOSPHORUS

2.10.3.1.1. *Principle*

The weighed sample (>30 μg) is digested in an open vessel with a perchloric acid–sulfuric acid mixture. The phosphate is determined spectrophotometrically as molybdovanado phosphate.

2.10.3.1.2. *Equipment*

Sample holder for weighing solid samples. Cf. Fig. 4*d*.

1-ml volumetric flask. Cf. Fig. 65. The little volumetric flask, with a ⸸ 7 joint, is calibrated by introducing exactly 1 ml water from a calibrated pipet and marking the neck.

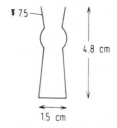

⸸ 75

4.8 cm

1.5 cm

Figure 65. 1-ml volumetric flask.

Heating block with thermometer. 250–300° (cf. Fig. 29).

Spectrophotometer with provision for microcuvets

2.10.3.1.3. *Reagents*

10% (g/v) aqueous ammonium molybdate solution. Kept in a polyethylene pipet bottle with a 100-μl pipet (cf. p. 22).

1% ammonium vanadate solution (g/v) in 2 N nitric acid. 1 g ammonium vanadate a.r. is dissolved in about 50 ml water and 12.8 ml concentrated nitric acid and the solution is diluted to 100 ml. Kept in a polyethylene pipet bottle with a 50-ml pipet.

70% perchloric acid a.r. Kept in a glass pipet bottle with a 5-μl pipet.

Sulfuric acid a.r. sp. gr. $= 1.84$. Kept in a pipet bottle with a 30-μl pipet.

Phosphate standard solution with 1 μg phosphorus per 10 μl. 0.4393 g potassium hydrogen orthophosphate (KH_2PO_4 after Sörensen), dried *in vacuo* at 110°, is dissolved in a 1000-ml volumetric flask and diluted to volume with double-distilled water.

2.10.3.1.4. *Establishing the Calibration Curve for 0–20 μg Phosphorus*

0,25,50,75,100,125,150, and 200 μl phosphate standard solution are measured with a 500-μl micro piston buret into 1-ml volumetric flasks, and 30 μl concentrated sulfuric acid and 5 μl 70% perchloric acid are added to each. The flasks are heated on the heating block at 250–300° protected from dust. After perchloric acid fumes appear, heating is continued for 30 sec. The flasks are removed from the heating block and cooled, and to each is added 500 μl water from a micro piston buret, 50 μl ammonium vanadate solution, and 100 μl ammonium molybdate solution; then the flasks are filled to the 1-ml mark with water. The flasks are closed with polyethylene stoppers and shaken, and after 15 min the absorbance at 430 nm is measured in a microcuvet. The calibration curve is linear.

2.10.3.1.5. *Decomposition and Determination Procedure*

Solid samples are weighed by difference into the digestion volumetric flasks by means of the sample holders (cf. Fig. 4*d*):

Liquid samples are absorbed onto 5-mm cotton threads before weighing.

30 μl concentrated sulfuric acid and 5 μl 70% perchloric acid are added to the weighed sample and heated at 250–300° on the heating block until the perchloric acid fumes. After 30 sec the flask is removed from the heating block, cooled, and the procedure continued as described in the previous section.

A blank value determination is necessary for each series of samples.

2.10.3.2. METHOD OF BELCHER, MACDONALD, PHANG, AND WEST (2) FOR MORE THAN 2 μg PHOSPHORUS

2.10.3.2.1. *Principle*

The weighed sample is decomposed in the oxygen flask, the phosphate precipitated as quinoline phosphomolybdate, and the precipitate separated and titrated acidimetrically.

2.10.3.2.2. *Equipment*

Combustion flask. Cf. p. 42 ff. The platinum gauze should be formed into a little basket to prevent the sample from touching the gauze directly and thus having too much heat conducted off during the decomposition.

Filtration apparatus. Cf. Fig. 44, Sect. 2.5.3.1.5.3. The paper filter mass is supported by a perforated sheet of polyethylene.

Titration apparatus. Cf. Fig. 12, Sect. 1.2.4.2.

Dropping pipets. About 10-μl drop volume.

Three 500-μl micro piston burets. Cf. p. 24.

2.10.3.2.3. *Reagents*

Sodium molybdate solution. 15 g molybdenum trioxide a.r., 3 g sodium hydroxide a.r., and 50 ml distilled water are heated for 30 min. After the solution has been filtered, 46 ml concentrated hydrochloric acid a.r. and 2 drops 30% hydrogen peroxide are added. Kept in a polyethylene bottle.

Quinoline hydrochloride solution. 2.8 ml freshly distilled quinoline is dissolved in 60 ml hydrochloric acid (1:1). Kept in a polyethylene bottle.

Ca. 0.5 N sodium hydroxide solution. Sodium hydroxide a.r. dissolved in distilled water.

Saturated bromine water

Saturated aqueous boric acid solution

Ca. 1 N hydrochloric acid a.r.

0.02 N sodium hydroxide solution of known titer. Standardized against potassium hydrogen phthalate.

0.02 N hydrochloric acid of known titer. Standardized against 0.02 *N* NaOH.

Polyethylene film. Cf. p. 43.

Strips of ash-free filter paper. 10 x 2 mm. For ignition.

5% citric acid solution. Prepared fresh daily.

2.10.3.2.4. *Decomposition and Determination Procedure*

The weighed sample (30–70 μg) is wrapped in a piece (15 x 15 mm) of polyethylene film (cf. Sect. 2.5.3.1.4). A strip of ash-free filter paper (10 x 2 mm) is used as fuse. (A linen thread has not proved satisfactory as fuse here.)

After 100 μl 0.5 *N* sodium hydroxide solution and 100 μl saturated bromine water have been introduced into the decomposition flask, it is filled with oxygen (cf. Sect. 2.5.3.1.4) and the sample is ignited. Immediately after the decomposition the solution is distributed over the lower half of the wall by rotation of the flask. After 30 min the wall is wet again and the flask is let stand for another 60 min. It is then opened and the stopper, sample holder, and wall are rinsed with 1.5 ml water from a piston syringe (cf. Sect. 2.5.3.1.4). The flask is let stand 15 min to allow the water to drain.

If fluorine is suspected in the sample, 15 μl boric acid solution is added and the solution is heated to boiling. The sample holder is rinsed with 100 μl 1 *N* hydrochloric acid and then with a drop of water. The solution is heated on a boiling water bath until the greater part of the bromine has been removed.

Now 100 μl citric acid solution [60] is added, the mixture (about 1.5 ml) is heated to 80–90° on a water bath, and 60 μl sodium molybdate solution and 40 μl quinoline solution are added. The flask is shaken vigorously and then left standing on the water bath 1 min more. After the solution has cooled it is filtered through a paper filter with the filtration apparatus (Fig. 44) and the flask is rinsed three times with about 1–2 ml water from a wash bottle provided with a fine tip; the precipitate is washed until the wash water remains neutral. The filter and precipitate are transferred completely into the decomposition flask with some water, and the flask, provided with a stirring bar, is put into the titration apparatus. Pure nitrogen is introduced into the suspension from a fine inlet capillary. The nitrogen flow should not be interrupted during the subsequent titration. A known amount (400 μl usually suffices) of accurately standardized 0.02 N sodium hydroxide solution is now added from a 500-μl piston buret and the solution is stirred magnetically until the precipitate has dissolved completely. After the wall has been rinsed with some water, the solution is back-titrated against phenolphthalein with 0.02 N hydrochloric acid from a 500-μl piston buret. The blank value determinations must be performed under identical conditions. 1 μl 0.02 N sodium hydroxide solution corresponds to 23.8 ng phosphorus.

2.10.3.3.   PROCEDURES FOR 0.5–5 μg AND 0.1–1 μg PHOSPHORUS (3)

2.10.3.3.1.   *Principle*

The sample, measured out by solution partition, is digested with perchloric acid–nitric acid in an open quartz tube and the phosphate determined spectrophotometrically via molybdenum blue.

---

[60] This is added to prevent interference from silicic acid from the material of the vessel. The amount of citric acid must be adhered to accurately. With new combustion flasks, larger amounts of silicic acid may occasionally be dissolved, which are then not adequately masked by citric acid. In this case several blank combustions should be conducted before the determination to condition the flask.

2.10.3.3.2. *Equipment*

Quartz test tubes. 120 mm length, 10 mm diameter, with spherical distention below the $\overline{\$}$ 7.5 joint and with a 3-ml calibrating mark (Fig. 66). Fitting polyethylene or Teflon stoppers.

Aluminum heating block. With holes for the test tubes and a thermometer. The holes should be of such a depth that only the lower quarter of the digestion tubes is heated.

Glove box. Of PVC, through which filtered air can be drawn; or evaporation apparatus (cf. Fig. 29).

Pipetting apparatus. Cf. Sect. 1.2.4.1, Fig. 10.

Spectrophotometer. With 0.5-, 1-, or 2-cm semimicrocuvets.

2.10.3.3.3. *Reagents*

70% perchloric acid a.r. Distilled with a quartz still; <0.1 μg/ml phosphate.

$HClO_4$–HNO_3 digestion acid mixture (1:2). Cf. Sect. 2.10.3.2.3.

Ca. 1 N sodium hydroxide solution a.r. Kept in a polyethylene bottle.

Saturated aqueous 2,4-dinitrophenol solution. In a polyethylene pipet bottle with a 50-μl pipet.

Phosphate standard solution. 43.9_3 mg KH_2PO_4 after Sörensen, dried *in vacuo* at 110°, is dissolved in 10 ml 1 N sulfuric acid and diluted with water to 1000 ml. The solution is stable for several weeks in a polyethylene bottle (cf. Sect. 1.2.6). 10 μl corresponds to 100.0 ηg phosphorus.

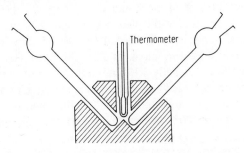

Figure 66. Quartz digestion vessels in heating block.

Molybdate reagent. 685 mg sodium molybdate (Na_2MO_4 · $2H_2O$) a.r. and 40.0 mg hydrazine sulfate a.r. are dissolved in 10 ml water in a 100-ml volumetric flask. While cooling with ice and stirring, 10 ml concentrated sulfuric acid a.r. and an additional 50 ml water are added. When the solution is again at room temperature it is diluted to volume with water. The yellowish brown solution is stable for about 3 weeks in a tightly closed brown $\mathbb{F}$ glass bottle in a refrigerator.

Methyl isobutyl ketone (*MIBK*) a.r. Freshly distilled; b.p.: 116.9°.

Double-distilled water from a quartz still is used for the preparation of all aqueous solutions.

2.10.3.3.4. *Establishing the Calibration Curves*

0.5–5 µg phosphorus. Phosphate standard solution, corresponding to 0.5–5 µg phosphorus, is measured into the calibrated quartz digestion tubes with a 500-µl micro piston buret and, after addition of 2 ml double-distilled water and 200 µl molybdate reagent (blood sugar pipet), heated for 30±5 min in a boiling water bath or a heating block at 98±2°. After the solution has cooled to room temperature it is diluted with double-distilled water to the 3-ml mark and transferred with the pipetting apparatus into 1-cm semi-microcuvets. The absorbance is measured within 60 min against an analogously prepared phosphate-free solution at 825 nm. The calibration curve is linear.[61]

0.1–1 µg phosphorus. Phosphate standard solution corresponding to 0.1–1 µg phosphorus is measured into the quartz digestion tubes with a 100-µl micro piston buret [e.g., buret part of the "Spinco" microtitrator of Beckman Co. (cf. p. 27 ff)] and, after addition of 2 ml double-distilled water and 200 µl molybdate reagent (blood sugar pipet), heated for 30±5 min in a boiling water bath or in a heating block at 98±2°. After this has cooled to 20° (thermostat), exactly 1.00 ml methyl isobutyl ketone at the same temperature is added and the vessel is closed with a polyethylene or Teflon stopper and shaken for 2 min. Then the phases are allowed to clear at 20° (thermostat) and the organic phase is

---

[61] Since the molybdate reagent has only limited stability, the calibration curve must be checked from time to time.

transferred by means of the pipetting apparatus into 1- or 2-cm semimicrocuvets, which are closed immediately and put into the dark. The absorbance at 780 nm is measured within 15 min at a 1–2° higher cell temperature against a phosphate-free reference solution. The calibration curve is linear; it must be checked occasionally, especially when new reagents are used.

2.10.3.3.5. Decomposition and Determination Procedure

Samples over 30 μg are weighed into the digestion vessel directly by difference (cf. Sect. 1.2.2.3). If the weighing is by solution partition (cf. Sect. 1.2.4.4), the residue should be localized on a small spot on the floor of the vessel. Large volumes of solution are transferred as described in Sect. 1.2.5.

100 μl perchloric acid–nitric acid mixture is added and heated with refluxing for about 40 min at not over 190° in the heating block. The perchloric acid must never be evaporated completely. The remainder is taken up in 500 μl double-distilled water, again heated to boiling for 3–4 min, and neutralized against 2,4-dinitrophenol (50 μl) with 1 N sodium hydroxide until a faint yellow color appears (rubber bulb pipet). After addition of double-distilled water to a total volume of about 2 ml. 200 μl molybdate reagent is added (blood sugar pipet) and the procedure is continued as described in the previous section.

2.10.4. REFERENCES

1. W. J. Kirsten and M. E. Carlsson, *Microchem. J.*, **4**, 3 (1960).
2. R. Belcher, A. M. G. Macdonald, S. E. Phang, and T. S. West, *J. Chem. Soc.*, **1965**, 2044.
3. G. Tölg, *Z. Anal. Chem.*, **194**, 20 (1963).
4. T. Salvage and J. P. Dixon, *Analyst*, **90**, 24 (1965).
5. L. J. F. Böttcher, C. M. van Gent, and C. Pries, *Anal. Chim. Acta*, **24**, 203 (1961).
6. H. N. Wilson, *Analyst*, **76**, 65 (1951).
7. R. Belcher and A. M. G. Macdonald, *Talanta*, **1**, 185 (1958).
8. R. Zinzadze, *Ind. Eng. Chem., Anal. Ed.*, **7**, 227 (1935).
9. F. L. Hahn and R. Luckhaus, *Z. Anal. Chem.*, **149**, 172 (1956).
10. R. P. A. Sims, *Analyst*, **86**, 584 (1961).
11. W. Geilmann and G. Tölg, *Glastech. Ber.*, **33**, 376 (1960).

12. F. Umland and G. Wünsch, *Z. Anal. Chem.*, **213**, 186 (1965).
13. R. A. Chalmers and D. A. Thomson, *Anal. Chim. Acta*, **18**, 575 (1958).

2.11. ARSENIC

2.11.1. DECOMPOSITION, GENERAL

Until now, only samples of over 30 μg have been decomposed in the oxygen flask for arsenic determination (1). Since arsenic alloys with the platinum of the usual sample holder, the sample, wrapped in polyethylene film, must be wedged into a quartz coil. To facilitate combustion, a small crystal of potassium nitrate may be added to the sample. Sodium hypobromite solution has proved satisfactory as absorbent.

2.11.2. DETERMINATION, GENERAL

Belcher, Macdonald, Phang, and West (1) found the spectrophotometric molydenum blue procedure of Morris and Calvery (2) suitable for the determination of 2–20 μg arsenic. Phosphate interferes.

2.11.3. PROCEDURE FOR ARSENIC DETERMINATION

2.11.3.1. PROCEDURE OF BELCHER, MACDONALD, PHANG, AND WEST FOR MORE THAN 2 μg ARSENIC (1)

2.11.3.1.1. *Principle*

The sample (30–70 μg) is decomposed in the oxygen flask, the arsenate is converted to molybdo arsenic acid, which is reduced by hydrazine to molybdenum blue, and this is determined spectrophotometrically.

2.11.3.1.2. *Equipment*

Combustion flask. Cf. p. 42 ff. With a sample holder of quartz instead of platinum wire. The holder consists of a thin quartz rod (40–50 mm length, 1 mm diameter) fused at one end to the stopper and wound into a spiral of 3–4 turns at the other end.

Spectrophotometer. With 1–cm cuvets.

10–ml volumetric flask

2.11.3.1.3. *Reagents*

Ca. 0.5 N sodium hydroxide solution. Sodium hydroxide a.r. dissolved in distilled water.

Saturated bromine water

Arsenic standard solution. 0.132 g well dried arsenic trioxide a.r. is dissolved in 4 ml 0.5 N sodium hydroxide solution in a 100-ml volumetric flask. The solution is diluted with water, weakly acidified with hydrochloric acid, and diluted to volume with water. The working solution is diluted 1:10; it contains 100 μg arsenic/ml.

Ammonium molybdate solution. 5 g ammonium molybdate a.r. is dissolved in 1000 ml 2 N sulfuric acid.

Hydrazine sulfate solution. Aqueous 0.025% (g/v) solution, which must be prepared fresh daily.

Polyethylene film. Cf. p. 43.

Strips of ashless filter paper. 10 x 2 mm; for ignition.

Ca. 1 N sulfuric acid

2.11.3.1.4. *Decomposition and Determination Procedure*

The sample (30–70 μg) is wrapped in a piece of polyethylene film (cf. Sect. 2.5.3.1.4) and the packet is wedged with a paper strip fuse in the spiral of the quartz holder. 100 μl 0.5 N sodium hydroxide solution and 100 μl bromine water are added, the decomposition flask is filled with oxygen, and the sample is burned (cf. Sect. 2.5.3.1.4). At the end of the combustion the flask is immediately shaken to wet the lower half of the wall with absorbent solution. After 30 min it is shaken again and is then left standing for 60 min. Then the flask is opened, 100 μl 1 N sulfuric acid is added to the absorbent solution, and the solution is boiled until all the bromine has been driven off. 4 ml water and 2 ml ammonium molybdate solution are added with pipets, rinsing the quartz holder and vessel wall. After addition of 2.5 ml hydrazine sulfate solution, the liquid is mixed and heated for 10 min on a water bath at 90–100°. The solution is cooled to room temperature and transferred to a 10-ml volumetric flask. For complete transfer, the decomposition flask is rinsed several times with water (about 2 ml total) and the volumetric flask is filled to the mark with the rinse water. The absorbance of the colored solution is measured at 840 nm in 10-mm cuvets against an analogously prepared arsenic-free reference solution. The color does not change within 24 hr.

The procedure for establishing a calibration curve is similar, except that known amounts of arsenic standard solution are introduced corresponding to 2–20 μg arsenic and an arsenic- and phosphorus-free substance is burned (e.g., benzoic acid).

2.11.4. REFERENCES

1. R. Belcher, A. M. G. Macdonald, S. E. Phang, and T. S. West, *J. Chem. Soc.*, **1965**, 2044.
2. H. J. Morris and H. O. Calvery, *Ind. Eng. Chem., Anal. Ed.*, **9**, 447 (1937).

196 PROCEDURES FOR DETERMINATION OF THE ELEMENTS

ILLUSTRATION CREDITS

Fig. 1. F. Hecht and M. K. Zacherl, *Handbuch der Mikrochemischen Methoden,* Vol. I, Part 2. Waagen und Wägungen von Benedetti-Pichler, A.A., Springer, Wien, 1951, p. 153, Figs. 52*a* and 52*b*.

Fig. 2. R. Belcher, *Submicro Methods of Organic Analysis,* Elsevier, Amsterdam, 1966, p. 10, Fig. 2.3.

Fig. 6. K. Ballschmiter and G. Tölg, *Z. Anal. Chem.,* **203**, 20 (1964), Fig. 5.

Fig. 10. B. Morsches and G. Tölg, *Z. Anal. Chem.,* **200**, 20 (1964), Fig. 9.

Fig. 20. K. Ballschmiter and G. Tölg, *Z. Anal. Chem.,* **203**, 20 (1964), Fig. 1.

Fig. 23. G. Tölg, *Z. Anal. Chem.,* **194**, 20 (1963), Fig. 1.

Fig. 26. W. J. Kirsten, *Microchem. J.,* **7**, 34 (1963), Fig. 1.

Figs. 30, 31, and 32. P. Gouverneur, H. C. E. van Leuven, R. Belcher, and A. M. G. Macdonald, *Anal. Chim. Acta,* **33**, 360 (1965), Figs. 2, 3, and 4.

Fig. 33. K. Ballschmiter and G. Tölg, *Z. Anal. Chem.,* **203**, 20 (1964), Fig. 6.

Fig. 34. K. Ballschmiter and G. Tölg, *Z. Anal. Chem.,* **203**, 20 (1964), Fig. 7.

Fig. 35. G. Tölg and K. Ballschmiter, *Microchem. J.,* **9**, 257 (1965), Fig. 2.

Figs. 36 and 37. A. Campiglio, *Il Farmaco, Ed. Sci.,* **19**, 385 (1964), Figs. 1 and 2.

Figs. 38, 39, and 40. W. J. Kirsten, and K. Hozumi, *Mikrochim. Acta,* **1962**, 777, Figs. 1, 2, and 3.

Fig. 41. W. H. List and G. Tölg, *Z. Anal. Chem.,* **226**, 127 (1967), Fig. 2.

Figs. 45, 47, 48, 49, and 50. G. Tölg, *Z. Anal. Chem.,* **194**, 20 (1963), Figs. 6, 2, 3, 4, and 5.

Figs. 52, 53, and 54. G. Schwab and G. Tölg, *Z. Anal. Chem.,* **205**, 29, (1964), Figs. 2, 3, and 6.

Figs. 55 and 56. W. H. List and G. Tölg, *Z. Anal. Chem.,* **226**, 127 (1967), Figs. 3 and 4.

Figs. 57, 58, 59, 61, 62, and 63. B. Morsches and G. Tölg, *Z. Anal. Chem.,* **200**, 20 (1964), Figs. 1, 4, 5, 7, 8, and 2.

INDEX

197